Cancer and You

Cancer and You
How to stack the odds in your favour

Malcolm H. Goyns
School of Sciences
University of Sunderland
UK

harwood academic publishers
Australia • Canada • China • France • Germany • India
Japan • Luxembourg • Malaysia • The Netherlands
Russia • Singapore • Switzerland

Amsteldijk 166
1st Floor
1079 LH Amsterdam
The Netherlands

British Library Cataloguing in Publication Data

Goyns, Malcolm H.
   Cancer and you : how to stack the odds in your favour
   1. Cancer – Popular works 2. Cancer – Prevention – Popular
works
   I. Title
   616.9′94

   ISBN: 90-5702-445-4

Cover illustration provided by Simon Perrins.

Dedicated to my parents

CONTENTS

PREFACE

Cancer is one of the major causes of death in our society, accounting for perhaps 25% of all mortalities. Although many more people die of heart disease, the word cancer holds a special dread for most people. This attitude is of particular concern, because a fear of cancer could lead to an individual delaying contacting their doctor and thus losing the opportunity of receiving treatment which might have resulted in a cure.

During the past decades there have been immense scientific and clinical research efforts dedicated to understanding the nature of malignant disease. It has been in the last decade, however, with the application of molecular biology techniques, that the major advances in our knowledge have occurred. This increased understanding of the genetics of cancer has provided an important insight as to why certain environmental factors appear to be associated with a much increased risk of developing cancer.

There is still a great deal to be discovered about cancer and I am sure that, as soon as this book is published, parts of it will be superseded by new information. Nevertheless, it is now possible for anyone to make value judgements about their attitude to cancer and their lifestyle; judgements that could significantly reduce their risk of developing and dying from the disease. The purpose of this book is therefore twofold. Firstly I have attempted to explain what cancer is, in a way that will hopefully demystify it for the reader, and will encourage them to be less frightened of confronting the disease if it does occur. Secondly, I have discussed the main causes of cancer, as far as we know them today, and have indicated how changes in lifestyle could significantly lessen the risk of developing cancer.

In writing this book, I would like to thank Professor Barry Hancock, Dr John Goepel and Dr Kilian Mellon for their comments on the chapters dealing with cancer diagnosis and treatment. I would also like to thank the many non-specialists whose criticisms were particularly valuable in hopefully making this an easy book to understand for the general reader. Finally I thank Professor Hancock, Dr Goepel and Dr Mohammed Alqahtani for providing figures for the colour and half-tone plates.

Section I

Everything you wanted to
know about cancer but were
afraid to ask

Chapter 1
THE CANCER PROBLEM

Introduction

After heart disease, cancer is the most common cause of death. It has been estimated that in industrial countries approximately one in three people will develop cancer at some time during their life, and probably one in four will die from the disease. In the United Kingdom alone it is estimated that 300,000 new cases of cancer are diagnosed and over 150,000 people die from the disease each year. World wide this is translated as 10 million new cases and 7 million deaths from cancer each year. This has important consequences both at government level, where the costs of treatment may represent almost 10% of the health service budget, to that of the individual, where cancer can result in a personal and family tragedy. **Cancer incidence**, that is the number of cases which occur in a population, is always a higher figure than **cancer mortality**, that is the number of deaths due to cancer in a population. This is because a number of patients with cancer are cured or, in some cases, cancer patients die from other causes. However, the majority of patients still die from malignant disease and it therefore remains a major health issue.

In the face of this enormous problem there is some surprisingly good news in that there is now clear evidence that the risk of developing cancer can be greatly reduced by adopting certain lifestyle options. Cancer is a difficult disease to analyse because its causes are very complicated, and also because, as it may take so long to develop, the important causative events may no longer be present when the disease is finally noticed. Nevertheless, a great deal of research has been carried out during the past few decades to identify key factors that either promote or restrict the development of cancer. As a result of these studies it has been estimated that environmental factors may be primarily responsible for up to 80% of all malignant disease. If this is true, then it represents a wonderful opportunity to reduce cancer incidence and hence relegate it to the category of infrequent diseases.

In this book the latest information is presented concerning the factors thought to be involved in causing cancer, showing how these are often factors associated with lifestyle. Sensible approaches that can be adopted to stack the odds in your favour of avoiding cancer are then indicated.

Although an increasing number of cures are being recorded with the advent of new therapeutic approaches, preventing the disease from occurring in the first place is still the best option.

A History of Cancer

Cancer is not a new disease as some people have thought. It has been known for thousands of years in human populations and evidence for it has even been found in the fossilised bones of prehistoric animals. Written records dating back to ancient Egypt and Greece describe diseases that were almost certainly cancers. Hippocrates was possibly the first to use the word *onkos* as a term to describe a swelling. In its earliest usage this was probably not used exclusively to describe a tumour, but it has since given rise to the term **oncology**, the study of cancer. In fact the word, cancer, which meant crab in the ancient world, was used as a description of breast cancer. This was because the radiating blood vessels from a central cancer mass had the appearance of a crab to the doctors of that time. There are now several terms that are commonly used to describe a cancer including **tumour**, **neoplasm** and **malignancy**. These tend to be used interchangeably. Terms such as **carcinoma**, **sarcoma** and **leukaemia** describe particular types of cancer, as is discussed in more detail in Chapter 4.

The difficulties in treating cancer were also noted in ancient times and the importance of attempting treatment early in the progress of the disease was mentioned by the second century doctor, Galen. He stated that small tumours could be cured by surgery but not the larger, more advanced ones. This is as true today as it was then. There were a number of reports of tumours in the following centuries, but the study of cancer did not really expand until the eighteenth century. This was primarily due to the advent of post-mortems when tumours in a variety of organs could be described. Morgagni published an influential treatise describing cancers in different organs in 1761. A few years later Potts described a possible causative link between soot and cancers in chimney sweeps; thus linking the environment with the onset of malignant disease. In 1838 Mueller described the cellular nature of tumours, which was an important observation that later suggested that tumours may arise by the abnormal proliferation of cells.

Up to and including the nineteenth century the main way of treating cancer was by **surgery**. A number of other therapies were tried, but all were largely without effect. The twentieth century has seen two major advances in anti-cancer treatment. The first of these is **radiotherapy** which was initially developed in the early part of the century. At first this type of approach was used to control the disease and reduce the tumour mass. However, in the 1950s it was reported that radiotherapy could produce a cure in a form of cancer called Hodgkin's disease. The twentieth century also saw the development of **chemotherapy**. The first chemicals which were found to have an effect on cancers were based on mustard gas, the gas that killed so many during World War I. The improvement of chemotherapeutic agents has continued to the present, with the development of drugs that are toxic to malignant cells but which have fewer toxic side effects on normal cells.

Both radiotherapy and chemotherapy offer opportunities for controlling and in some cases curing cancer. However, their use is a balancing act. On the one hand they are used in the most effective way for the destruction of unwanted tumours but, at the same time, they should cause the least damage to the normal tissues of the patient. There are after all some very effective ways of destroying all cancers; cyanide is a good example, but unfortunately it will also kill the patient. So the search continues to find a treatment which will efficiently kill all cancers, yet leave the patient relatively unharmed.

Cancer Incidence

It has been suggested that cancer is mainly a disease of modern industrialised society, but this is difficult to prove. Certainly it seems to be less common in primitive hunter gather communities, but such people tend to have a much shorter lifespan than individuals in developed countries, and cancer is primarily a disease of middle to old age. Furthermore, it is likely that in older records of deaths, such as those that exist in England, cancers were often misdiagnosed and so the apparent incidence was much lower than in reality. This debate has continued during the twentieth century. There does seem however to be an increasing body of evidence which indicates that cancer rates may have increased throughout the last century.

Mortality figures for cancer have also been available for the past century, but there is still debate as to what apparent changes in cancer trends really mean and whether progress being made in the treatment of the disease can be monitored in this way. After all, a fall in mortality caused by one type of cancer may be due to improvements in treatment, or could be due to earlier detection which often leads to a greater chance of a cure, or might even be due to changing environmental factors affecting the incidence of the disease.

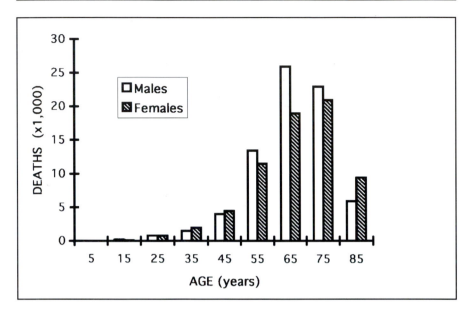

Figure 1.1 Annual number of cancer deaths in the United Kingdom in different age groups.

One thing that is not in dispute is that cancer incidence generally increases with age. Figure 1.1 summarises the deaths from cancer as a function of age and clearly shows that there is an increasing number of cases in middle to old age. There are relatively few cases of cancer under 40 years of age and although cancer in children is a very emotive subject, it is in fact relatively uncommon. In developed countries it has been estimated that 1 in 600 children under the age of 15 years is affected by cancer. This still means though, that in the United Kingdom alone there are over a thousand new cases reported in children each year.

Cancer can affect different parts of the body and there is considerable variation as to which organs are most affected. When both sexes are considered together, the most commonly occurring cancers are lung, stomach, breast, colon and rectum, mouth and throat, liver, and cervix. However, the most common cancers to cause death are slightly different. The main ones being lung, stomach, liver, colon and rectum, oesophagus, and breast. These observations are summarised in more detail in Table 1.1. This information indicates that not all cancers are the same and that some can be treated more effectively than others. Why this should be is considered in more detail in later chapters.

Table 1.1 World Health Organisation estimates of new cases of cancer, and deaths from cancer, world-wide in 1996.

	MEN			WOMEN	
Cancer type	New cases (thousands)	Deaths (thousands)	Cancer type	New cases (thousands)	Deaths (thousands)
Lung	988	878	Breast	910	390
Stomach	634	518	Cervix	524	241
Colon and rectum	445	257	Colon and rectum	431	253
Prostate	400	204	Stomach	379	317
Mouth and throat	384	237	Lung	333	282
Liver	374	370	Mouth and throat	192	129
Oesophagus	320	305	Ovary	191	125
Bladder	236	107	Uterus	172	68
Other	1,531	1,043	Other	1,874	1,387
TOTAL	5,312	3,919		5,006	3,192

One of the most striking features revealed by studies of cancer incidence is the remarkable variation in the types of cancer found in different parts of the world. Minor variations in the structure of certain genes, which are known as genetic polymorphisms, are known to predispose individuals to develop cancer. However, the regional differences in cancer incidence do not seem to be due to racial differences of the indigenous populations, but rather to cultural and environmental differences. This again indicates the importance of lifestyle in influencing the risk of developing cancer.

A comparison of white populations in North America with oriental populations in China and Japan (see Table 1.2) demonstrates very significant differences in the incidences of some of the main types of cancer. Lung cancer is 50% higher in both males and females in North America than it is in Asia. Cancers of the colon and rectum are at least twice as common in North America. However, the largest difference is shown for breast and prostate cancers which are 3 to 15 times more common in the North American population. In contrast, the incidence of stomach cancer is approximately six times more common in Asia. At first sight these observations might indicate that there is a racial difference in the susceptibility to certain types of cancer, but studies of people of Japanese origin who have lived for many years in the USA have indicated

Table 1.2 Cancer incidence in different regions of the world (Rates per 100,000).

Country	Lung		Stomach		Colon & rectum		Prostate	Breast
	male	female	male	female	male	female	male	female
Canada	69	24	12	5	44	34	51	71
USA (white)	64	30	8	4	47	33	62	89
USA (Japanese)	35	18	30	14	55	40	33	73
Japan	42	12	74	33	27	16	7	22
China	53	18	52	22	18	16	2	21
England	75	19	22	9	38	25	25	63
Denmark	59	23	13	6	38	30	30	69
Germany	70	7	20	12	41	30	29	56
Slovakia	79	8	27	12	34	21	20	35
Poland	65	16	22	9	21	16	12	39
Belarus	56	5	47	20	18	13	9	25

that their patterns of cancer incidence are gradually approaching those of the white American population. They have, for example, half the incidence of stomach cancer found in Japan, but their incidences of breast and prostate cancers have increased by three and five times respectively compared to those observed in Japan. The conclusion is clear and that is that environmental factors are important in determining the onset of cancer.

Another comparison can be made of ethnically similar peoples who live in countries which have different economic conditions. The incidence rates in Table 1.2 demonstrate that lung, stomach, colon and rectum, prostate and breast cancers appear to occur at similar rates in England, Denmark and Germany. These are economically similar countries. However, comparison with the three East European countries of Slovakia, Poland and Belarus, indicates that stomach cancer is 50% more common in the latter countries. However, the incidences of breast and prostate cancers are only half those seen in West European countries. Again this appears to support the conclusion that environment is an important factor in cancer incidence.

The cancer incidence rates from developing countries may at first sight appear to be much lower than in industrialised countries. The World Health Organisation has surveyed cancer incidence and some of their findings are shown in Figures 1.2 to 1.6. These observations indicate that, for example, the incidence of a range of tumours is much lower in India than in Canada and Japan. However, as we have mentioned the average life expectancy is significantly lower in most developing countries and many people die from other causes before they reach

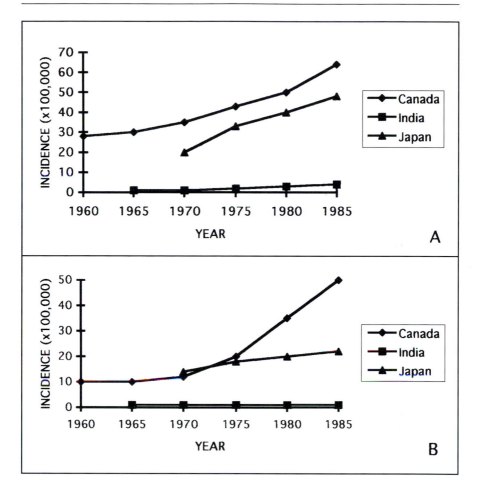

Figure 1.2 Incidence of lung cancer in Canada, India and Japan. (A) males; (B) females.

an age where cancer would become more common. There is also a problem relating to the correct recording of the cause of deaths in these countries. Studies in some central American countries have indicated that as the authorities become more proficient at diagnosing and recording the true cause of death, so the incidence rates of cancer appear to have increased.

As mentioned above the chances of surviving cancer depends on the type of cancer contracted. However, it has also become apparent that the mortality rates due to different types of cancer can vary from

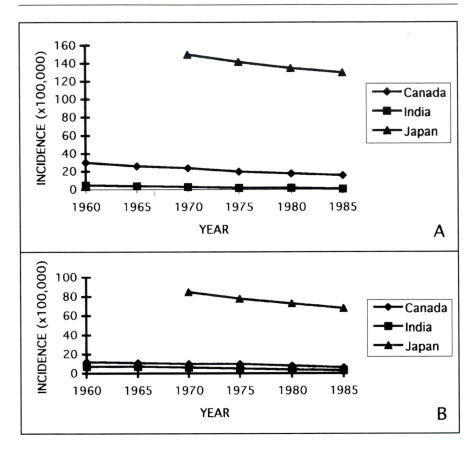

Figure 1.3 Incidence of stomach cancer in Canada, India and Japan. (A) males; (B) females.

population to population. In breast cancer it has been shown that the incidence to mortality ratio is 2.6 in the Netherlands but only 1.4 in Hungary. In other words a woman is almost twice as likely to survive breast cancer in the Netherlands than if she lives in Hungary. This effect has even been observed within countries where there are economic differences between two groups of the population; in the USA the incidence to mortality ratios for many cancers are more favourable for whites than blacks. This indicates that the prevalence of early diagnosis, access to better quality of treatment and aftercare all play a role in increasing survival from the disease.

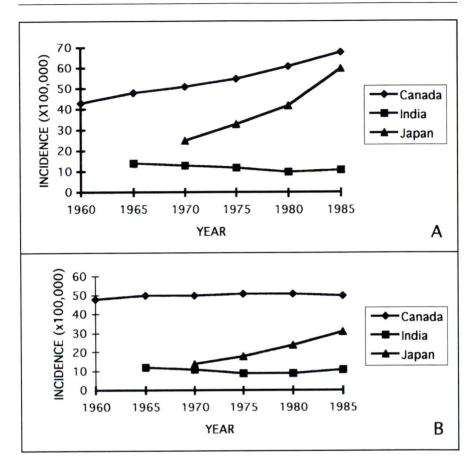

Figure 1.4 Incidence of colon and rectum cancer in Canada, India and Japan. (A) males; (B) females.

Future Cancer Trends

The incidence of cancer varies throughout the world, with certain types of cancer being more common in one region than another. There are, however, some general trends which indicate that certain cancers are steadily becoming more or less common with time. These trends can indicate what is likely to happen to cancer incidence rates in the future, at least in the short term. For example, in 1980 it was estimated that stomach cancer was the most common tumour for both sexes combined. However, less than 20 years on, it is now second most

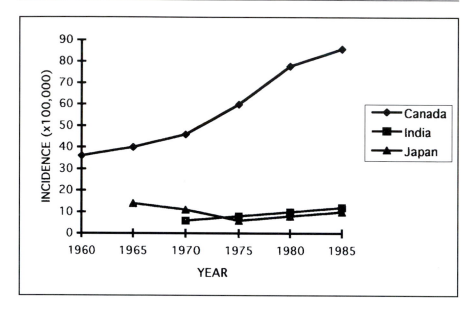

Figure 1.5   Incidence of prostate cancer in men in Canada, India and Japan.

common after lung cancer. The increase in lung cancer is continuing and is seen world-wide.

The rates of lung cancer have generally increased in all parts of the world and this varies with the proportion of people smoking, although the rates of lung cancer have begun to show a slight decline among men in Western countries with the heightened awareness of the risks of smoking. The increase in smoking in men in other regions of the world, and in women in Western countries, has led to dramatic increases in the incidence of lung cancer in these groups. For example, lung cancer incidence has increased by over 300% in Canadian women between 1970 and 1985 (Figure 1.2). Unless the dangers of tobacco usage are taken seriously by smokers, then this is a trend that will continue. This is of particular concern because lung cancer invariably leads to death within two to three years of diagnosis. Lung cancer is an aggressive malignancy which is not only resistant to most forms of treatment, but is a cancer where the early symptoms often go unnoticed. Therefore the opportunity for early treatment, leading to a possible cure, is missed.

By contrast the incidence rates for stomach cancer in both men and women have generally been decreasing (Figure 1.3). This trend has also been observed in all regions of the world, both in developed and developing countries. This has undoubtedly been influenced by changing

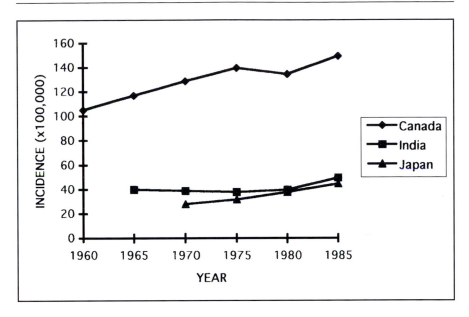

Figure 1.6 Incidence of breast cancer in women in Canada, India and Japan.

patterns of diet though the possibility remains that part of this trend may also be accounted for by earlier death from other causes, particularly heart disease caused by cigarette smoking.

Interestingly, tumours of the colon and rectum have become more common in recent years. In countries such as India there does not appear to have been a major change, but in the developed countries the increased incidence rates have been more obvious. This is particularly so in Japan where incidence rates in both men and women have increased by almost 300% in recent decades (Figure 1.4). At first sight this appears to be puzzling, because tumours of the stomach, colon and rectum are all likely to be susceptible to dietary factors. It appears therefore that certain components of the diet have a more profound effect on one type of tumour than another. Certainly different components of the diet can exhibit a protective action rather than pose a risk. This will be discussed in more detail in Chapter 10.

Although prostate tumours are generally becoming more common in all regions of the world, they are particularly common in industrialised countries (Figure 1.5). It has been suggested that the development of this tumour is influenced by diets that have a high meat and animal fat content, which are often significant components in the diets in Western countries.

In parallel to the increased incidence of prostate tumours in men, is the incidence of breast cancer in women. Although men can develop breast cancer, this is very much more common in females. It is a cancer which has shown increased incidence rates in many regions throughout the world over recent years (Figure 1.6). Again it is a malignancy which is much more common in developed countries and the increase in its incidence often parallels the rate of industrialisation. This does not necessarily relate to industrial pollution, but is more often due to other factors. Changes in diet caused by increased wealth in the society are important both in the type of food consumed and also because of the increased incidence of obesity. Cultural influences may also be important in industrialised countries, where women often delay pregnancy or have no children, and do not breastfeed their babies.

Although the degree by which cancer incidence rates alter can vary substantially from region to region, it appears that a number of general trends are consistent. As the genetic makeup of the different peoples of the world has not altered significantly in recent decades, these changes indicate that environmental factors are of great relevance. Observations such as these also demonstrate how remarkably quickly significant alterations in cancer incidence rates can occur. This conclusion should be regarded as a positive one for it indicates that by being aware of environmental risk factors and adopting appropriate lifestyles, a very significant reduction in cancer incidence could be achieved in a relatively short period of time; certainly within a generation.

Chapter 2
SOME BASIC BIOLOGY

Introduction

To understand what cancer is and why it occurs, it is necessary that the reader have a basic knowledge of biology. This need not be too arduous a task, as many of the concepts in biology are simple and straightforward. As is often the case in science, it is the terminology which mainly acts as a barrier to understanding. This chapter is therefore designed to keep such terminology to a minimum and hopefully will provide a gentle introduction into the fascinating world of **cells, chromosomes** and **genes**. Those more knowledgeable readers can probably ignore this chapter.

There are really two main lessons to be learnt. Firstly, that all living things, no matter how large or how small, are made up of tiny building blocks called cells. Microscopic organisms such as bacteria or yeast consist of a single cell, whereas humans, whales, daffodils and oak trees are made up of billions of cells. The latter are examples of multicellular organisms. In multicellular organisms the cells become specialised so that they can do different things. In a human, muscle cells can contract or relax (the basis of movement), white blood cells can attack invading bacteria (the basis of the immune response) and brain cells can conduct and process electrical impulses (the basis of communication including thought).

Secondly, the information to manufacture all of the **proteins** that make up a cell is stored in the hereditary material. This is true whether the proteins are structural proteins that maintain the shape of the cell, or enzymic proteins that catalyse the biochemical processes that control the activity of the cell. The instructions of how to make all types of proteins are stored in the hereditary material in a very long chemical called **DNA**. The DNA molecule is divided into regions which contain all of the information that enable the production of a protein, and this region is called a **gene**. These strings of genes are packaged into structures called **chromosomes**, which are stored in a part of the cell called the **nucleus**. In other words, each gene carries the information to make a particular type of protein.

In essence, cancer occurs because the DNA that makes up certain genes becomes altered. An alteration to a gene is called a **mutation**.

This in turn changes the amount or the nature of the proteins made from those genes, which may cause the cell to behave abnormally. Viewed in this way, the malignant transformation of cells seems so simple, but in practice it is complicated by the many different combinations of genetic changes that are able to produce a fully malignant cancer cell.

Cells — the Building Blocks of Life

One of the fundamental ideas in biology is that all living things are made up of cells. The first time the idea of a cell was suggested, was by Robert Hooke in the seventeenth century who noted, when studying pieces of cork under a microscope, that the wood appeared to be composed of little boxes. Shortly after that, Anton van Leeuwenhoek described a number of small single cell creatures after he had used a hand lens to study pond water. Gradually the concept of the cell as the basic building block of life developed.

Cells can be described as having a number of characteristics. They comprise all living things and represent the most elementary unit of life. All cells are derived from previously existing cells. This is usually the result of cell division, where one cell splits into two, grows and then splits into two again. However, it is also possible for a cell to arise by the fusion of a sperm and ovum (which are themselves specialised cells). Every cell is bounded by a **plasma membrane** which separates it from other cells and from the environment. Although many cells, especially in multicellular organisms, can have specialised functions, they all share certain basic biochemical activities. Finally, most cells are very small and visible only with the aid of a microscope. In general they are only one hundredth of a millimetre in diameter and tens of thousands of them can sit on the head of a pin. Some cells, such as bacteria, are very much smaller and some, such as nerve cells, are very much longer. The latter can stretch for several centimetres, though their diameters are still of microscopic proportions.

Although cells in a human body can do very different things and also look different when viewed down a microscope, they are all organised in a similar way. Figure 2.1 shows in a diagrammatic form all of the essential features of a typical human cell. The cell is bounded by a plasma membrane and it contains the material, known as **cytoplasm**, which makes up the cell. The membrane is composed of an extremely thin layer of fat and proteins. It serves not only to separate the cell from the surrounding environment, but also functions as a transport system

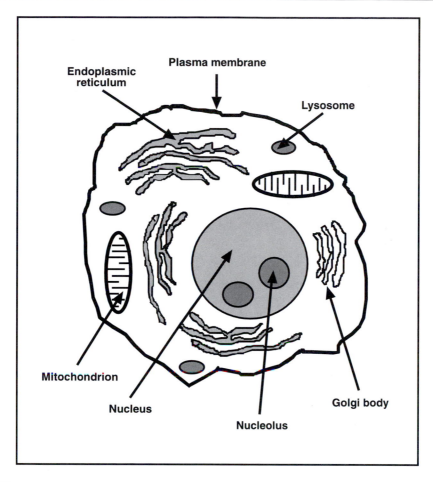

Figure 2.1 Diagrammatic representation of a typical human cell.

which controls the movement of chemicals in and out of the cell. The fatty layer is maintained by the action of protein enzymes within the cytoplasm and the membrane functions are controlled by proteins situated either in or on the membrane. So the whole structure and function of the membrane is dependent on proteins.

The cytoplasm is not an homogeneous bag of water or jelly, but is in fact composed of a complex array of membranous structures. These are the organelles and all have their special role within the cell. They include the **endoplasmic reticulum**, the **Golgi apparatus, lysosomes, mitochondria** and the **nucleus**.

The endoplasmic reticulum is a complex arrangement of membranes filling much of the cytoplasmic space. These membranes are often covered in small bodies called **ribosomes**. The ribosomes are the factories within the cell where new proteins are manufactured and there are thousands of them per cell. They are composed of an array of proteins and a chemical called **RNA**. This RNA molecule is similar to DNA (see below) and is of a special type that is only used to construct ribosomes. This has therefore been called ribosomal RNA or rRNA.

The Golgi apparatus is a specialised form of endoplasmic reticulum which is primarily found in cells which secrete proteins and appears to act as a storage site for these proteins. Lysosomes are used to store enzymes until they are needed. The mitochondria are the powerhouses of the cell. They are responsible for producing energy from sugars to supply the activity that occurs within the cell.

Finally there is the nucleus, which is arguably the most important part of the cell because it is the body that holds all of the genetic material. It is astonishing to think that all of the information needed to produce an entire human being is contained in the hereditary material stored in the nucleus of every single cell. The nucleus is bounded by a nuclear membrane and contains a number of structures called chromosomes which are packages of hereditary material. There are 46 chromosomes in the normal human cell; 22 pairs of somatic chromosomes and a pair of sex chromosomes (two X chromosomes in a female, and an X and a Y chromosome in a male). When viewed down a microscope the nucleus often also appears to contain one or more smaller bodies. These are the **nucleoli** which are the sites of rRNA synthesis.

In multicellular organisms, such as humans, the cells can become specialised during development of the embryo in the uterus. These cells can then perform a number of different functions in the body. This process of specialisation is called **differentiation** and occurs very early in embryonic development. The result is to produce all of the different types of cell that make up a fully formed individual. However, this process also occurs in an adult. A number of the cells in the body are lost through normal body function and have to be replaced. To enable the repopulation of specialised cells there are a number of immature **stem cells** which, when required, can begin to proliferate and differentiate into the required cell types.

An important distinction to make between all of the cells in the body is that some have the ability to grow and divide, so that they can either replace lost cells or build up their numbers, whereas other cells have become so specialised that they no longer have the ability to divide. Amongst the cells that can no longer divide are muscle cells and nerve

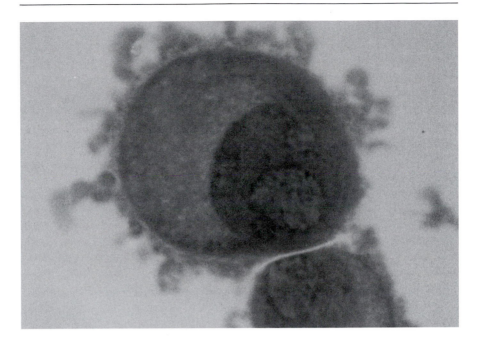

Figure 2.2 A human carcinoma cell. Note the large nucleus which fills the right side of the cell. Two nucleoli are visible within the nucleus.

cells. Cells that are able to divide include the epithelial cells and bone marrow cells. It is interesting to note therefore that malignant cells essentially arise from the latter group of cells; it is the cells which have retained the ability to divide that can be transformed into malignant cells, which then grow and proliferate out of control.

Epithelial cells give rise to the dead layer of skin on the outside of the body, which offers protection against environmental stresses. However, on the inside of the body the epithelial cells can act as a filter system. This means that the epithelial cells that line the gut can take nutrients from the food, and lower down the digestive tract can also expel unwanted chemicals. They can aid in the exchange of oxygen and carbon dioxide in the lungs, and regulate the passage of water and salts in the kidney. The fact that these linings of epithelial cells are the first contact the body has with the outside world means that they are the first cells to be exposed to cancer causing agents. It is not surprising therefore that the majority of cancers that arise in the body are cancers originating from **epithelial cells**, and these are called **carcinomas**. Figure 2.2 shows some human carcinoma cells which have been grown in culture. Although this type of photograph does not reveal all of the features that are described in the diagrammatic representation (Figure 2.1), they do have all of the characteristic features of a normal cell, including a distinctive nucleus with nucleoli.

The bone marrow cells are very active cells which produce the different types of blood cells. The **red blood cells** carry oxygen around the body but have lost the ability to divide and cannot become malignant. However, the **white blood cells** are involved in producing **antibodies** and in attacking invading bacteria and viruses. As part of this activity they can show an impressive capacity to divide and proliferate. The **bone marrow** cells, which include the immature white blood cells, can become transformed into a malignant state. When this happens, the increasing number of malignant bone marrow cells spill out into the blood and this is called a **leukaemia**. Other types of cancer are described in Chapter 4.

An understanding of the growth and division of cells is therefore central to understanding cancer. Figure 2.3 summarises the main events that occur during the growth and division of a normal cell. This is called the **cell cycle**. This process involves two main events. The first is the replication of all of the hereditary material so that there are two sets of hereditary information present in the cell, and the second is the orderly separation of these sets of information into two daughter cells. This process is accompanied by a general growth of the cell so that it becomes twice its original size. The original cell is usually in a state called the **G1 phase** of the **cell cycle** when it receives a signal to grow. This triggers the cell to move into the **S phase** or synthetic phase, where DNA (the chemical that makes up the hereditary material) is replicated. This is a remarkable process as the DNA molecules that make up each of the 46 chromosomes are faithfully copied to produce exact replicas of themselves, with all of their genetic information intact. During this time the cell is continuously growing. Once replication of the DNA is complete the cell enters the **G2 phase** of the cell cycle. Preparations now begin for cell division or **mitosis** (M phase). The first thing noticed in such cells is that the chromosomes condense so that they become clearly visible under the microscope. Each chromosome is duplicated and the original chromosome and its newly formed replica are held together at a region on the chromosome called the centromere. This often gives the chromosomes an X shape. The nuclear membrane then disappears and the chromosomes become attached to a protein array called the mitotic spindle. The chromosomes all line up along the equator of the cell and, after a signal is given, the spindle pulls each of the chromosomes away from its replica. As a result the original chromosome moves to one end

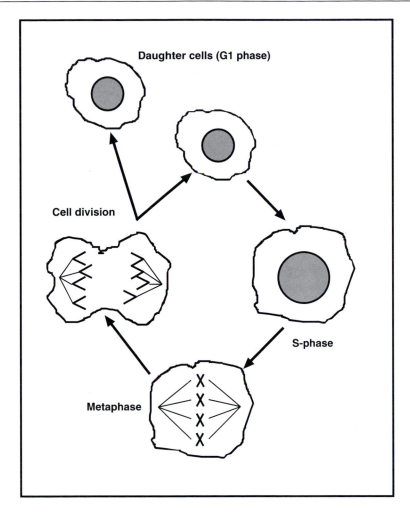

Figure 2.3 The cell cycle. A cell in the G1 phase grows and enters S-phase where DNA is synthesised. After S-phase the cell continues to grow until its nucleus disappears and the chromosomes condense and become visible. During metaphase the chromosomes align themselves along the mitotic spindle. The duplicate copies of chromosomes separate from one another and the cell begins to divide. Two daughter cells are formed, both of which are in G1.

of the cell and its replica to the other end. New nuclear membranes then form around each set of chromosomes and the cell divides to give rise two new daughter cells, which will now be in the G1 phase of the next cell cycle.

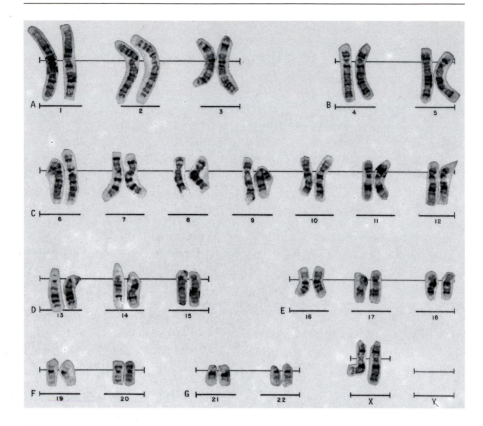

Figure 2.4 Karyotype of human chromosomes from a normal cell. The karyotype consists of 22 pairs of somatic chromosomes and two sex chromosomes. In this case the cell originated in a female and therefore there are two X chromosomes.

The chromosomes that become visible during mitosis can be photographed and then arranged according to size as a **karyotype**. An example of a karyotype is shown in Figure 2.4. This clearly shows the 22 pairs of somatic chromosomes and a pair of sex chromosomes. As the sex chromosomes in this case are both X chromosomes, we know that the cell that contained them came from a female. A cell from a male would have contained one X chromosome and one Y chromosome.

Many of the proteins that are involved in the mechanisms that control progress through the cell cycle have now been identified and we have a good understanding of how most of them interact in the cell. What is particularly interesting about these investigations is that a number of these regulatory proteins have also been found to be altered in cancer cells.

DNA, Genes and Chromosomes — The Hereditary Code

The cell is the basic unit of all life and it is clear that the structure, integrity and activity of each cell primarily depends on the action of proteins. Proteins lend structural support to the cell. Proteins catalyse all of the chemical reactions within the cell which define life. The instructions concerning what proteins to manufacture, when to manufacture them and in what cell type, is contained in the hereditary material. This is the information that is passed from one generation to the next, and from a cell to each daughter cell. The instructions to manufacture each protein in encoded by a unit of hereditary information called a gene, which is in turn composed of a chemical called DNA.

It was in the 1940s that bacterial genes were shown to be composed of DNA and it was not too long before this was also demonstrated to be true of animal and plant genes. In the 1950s, Francis Crick and James Watson demonstrated that DNA was composed of a long double stranded molecule. The best way to visualise what DNA looks like is to imagine two strings of beads twisted one around the other hence the name **double helix**. This double stranded structure is important because it explains how DNA can be replicated during the S-phase of the cell cycle and how messenger molecules can be formed to transfer the genetic information contained in the genes to the protein synthesising ribosomes (see below).

DNA is in fact a long chain made up of just four chemical units, called **nucleotide bases**. The names of these nucleotide bases are **adenosine**, **cytosine**, **guanine** and **thymidine**. These are often written in shorthand as just A, C, G or T. This is analogous to each string of beads being composed of beads of only four different colours. When this was first discovered, it was something of a puzzle, because it was difficult to see how just four different nucleotide bases could provide the information that could determine which of the 20 or so **amino acids** should be selected to make a protein. A protein is also a molecule that can be thought of as a string of beads, but in this case the beads are amino acids not nucleotide bases and there are 20 of them.

It was almost another 10 years, in the early 1960s, before it was proven that the nucleotide bases were arranged in triplets (called **codons**) and that each combination of three nucleotide bases indicated to the ribosome which amino acid should be attached to the growing protein that it was manufacturing. For example, the codon AAA codes for the amino acid phenylalanine. There are of course 64 possible combinations of three nucleotides, which means that some amino acids are coded for by more than one codon. Also, as there are more codons than necessary to code for the 20 amino acids, some of the codons are used to signal the end of making a protein. The signals coded for by these 64 codons has been called the **genetic code** (Table 2.1).

The genetic information that is stored in the DNA which makes up the chromosomes has in some way to be sent to the ribosomes, which manufacture proteins. To do this, a message is sent from any gene that is active in the cell. It is thought that 25% of the genes present in the chromosomes may be active in any cell at any time. As there are between c.80,000 and c.100,000 genes present in a human cell, this represents a very large body of information that has to be passed from the nucleus to the ribosomes.

It has been shown that the nucleotide bases can form weak links between one another, but always in a very specific manner. That is, adenosine always pairs with thymidine and cytosine always pairs with guanine, so that a nucleotide sequence of, for example, AACTGACGG on one strand of the DNA will be matched by a sequence of TTGACTGCC on the other complementary strand. Replication of the DNA during the cell cycle depends upon this complementary nature of the DNA double helix structure. During the S phase of the cell cycle the two strands of the DNA separate and each strand acts as a template for the synthesis of another complementary strand, resulting in the formation of two double helix structures, which are later segregated at mitosis. To synthesise a new DNA strand, free nucleotide bases in the nucleus pair weakly with the bases that make up each of the separated DNA strands, A always pairing with T and C always pairing with G. An enzyme called **DNA polymerase** is then responsible for binding together the individual nucleotide bases that are attracted to the separated DNA strands. In this way it ensures that the new strands are synthesised to give a perfect complementary copy (Figure 2.5).

A similar mechanism is used to produce the message which carries the genetic information from a gene and to the ribosomes. In this case the complementary DNA strands that form the double helix separate locally in the region containing the gene in question. An enzyme called **RNA polymerase** then synthesises a single stranded complementary copy of one of the DNA strands that comprises the gene. This is a **RNA** molecule and because it acts as a messenger molecule has been termed messenger RNA or **mRNA**.

The RNA molecules also comprise four different nucleotide bases but

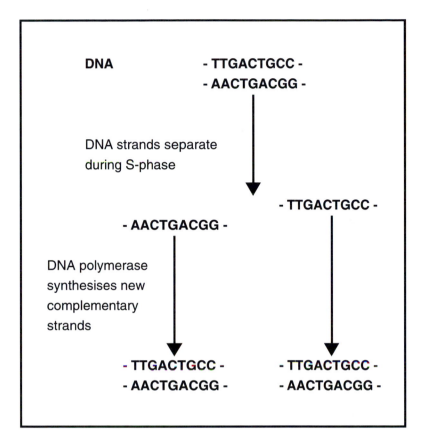

Figure 2.5 Key steps in the replication of DNA. These processes occur in the nucleus of the cell during S-phase.

instead of containing thymidine, they contain a different nucleotide called **uracil**, which is denoted as U. Therefore a sequence of nucleotide bases in DNA such as AACTGACGG will be transcribed into a mRNA molecule as UUGACUGCC (Figure 2.6).

The mRNA molecules migrate from the nucleus and become attached to a ribosome which then reads the instructions contained in the nucleotide sequence and synthesises an appropriate chain of amino acids. This means that a nucleotide sequence of AAC TGA CGG in the gene is eventually translated as an amino acid sequence of leucine-threonine-arginine in the protein (Table 2.1). This process results in the formation of a protein which can then carry out its role in the cell.

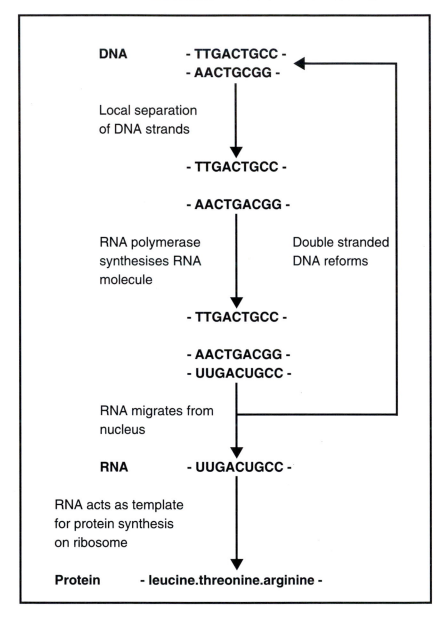

Figure 2.6 Key steps in the conversion of the genetic information held by a gene into protein. The separation of DNA strands and synthesis of RNA occurs in the nucleus of the cell. The RNA then migrates into the cytoplasm of the cell where it binds to a ribosomes associated with the endoplasmic reticulum and synthesis of protein occurs.

Table 2.1 The genetic code. The three nucleotide bases which comprise every possible codon in a mRNA molecule are shown along with the corresponding amino acid that they code for. STOP represents a codon which terminates protein synthesis.

First nucleotide	Middle nucleotide				Last nucleotide
	Uracil	Cytosine	Adenine	Guanine	
Uracil	Phenylalanine	Serine	Tyrosine	Cysteine	Uracil
	Phenylalanine	Serine	Tyrosine	Cysteine	Cytosine
	Leucine	Serine	STOP	STOP	Adenine
	Leucine	Serine	STOP	Tryptophan	Guanine
Cytosine	Leucine	Proline	Histidine	Arginine	Uracil
	Leucine	Proline	Histidine	Arginine	Cytosine
	Leucine	Proline	Glutamine	Arginine	Adenine
	Leucine	Proline	Glutamine	Arginine	Guanine
Adenine	Isoleucine	Threonine	Asparagine	Serine	Uracil
	Isoleucine	Threonine	Asparagine	Serine	Cytosine
	Isoleucine	Threonine	Lysine	Arginine	Adenine
	Methionine	Threonine	Lysine	Arginine	Guanine
Guanine	Valine	Alanine	Aspartic acid	Glycine	Uracil
	Valine	Alanine	Aspartic acid	Glycine	Cytosine
	Valine	Alanine	Glutamic acid	Glycine	Adenine
	Valine	Alanine	Glutamic acid	Glycine	Guanine

Mutations and Altered Proteins

In essence, a **mutation** is a change to the nucleotide sequence of a DNA molecule. When such a change occurs in a gene which codes for a protein then this can result in an alteration of the structure of that protein and possibly a change in its activity. Mutations can occur in many different ways. A number of environmental factors such as chemicals and physical agents (eg., **radiation**) can damage DNA. There are very reliable repair mechanisms present in human cells and this type of damage is usually corrected. However, if this damage is not repaired by the cell then the mutation becomes a permanent change in the DNA and can be inherited. If a mutation becomes established in a **germ cell**, that is a sperm or an ovum, then it can be passed on to future generations. If the mutation is established in a body cell, then it can be passed on to all subsequent daughter cells that are derived from that original cell.

There are different types of mutations. A mutation might change a single nucleotide base in a gene. In some cases this does not affect the amino acid in the corresponding protein because more than one codon often represents the same amino acid (Table 2.1). However, if a single nucleotide change does lead to a different amino acid being incorporated into the resulting protein, then this could have an effect on the normal activity of that protein. Such an altered protein may no longer function as it should, or it may begin to act in a novel and unexpected way. There is also the possibility that a mutation could convert a codon from one that codes for an amino acid to one that prematurely terminates synthesis of the protein. This again could have profound effects on the resulting truncated protein, which may either be inactive or exhibit abnormal activity.

Other forms of mutation, especially as a result of exposure to radiation, can lead to small pieces of DNA being deleted so that the gene is damaged to such an extent that it cannot produce a viable protein. It is also possible that the process of DNA replication itself may be affected so that DNA strands from different chromosomes become joined together. This produces abnormal chromosomes which are hybrids composed of two or more different chromosomes. This can result in the creation of chimeric genes which code for abnormal proteins with unusual activities.

Generally speaking mutations are either of limited significance, or are lethal and kill the cell. However, there is a group of mutations which affect a small group of genes which can have subtle but very influential effects on the cell. It is this group of mutations which form the basis of malignant transformation of a normal cell into a cancer cell.

So How do Cells Form a Tumour?

The human body represents a remarkable balancing act in which a myriad of influences are controlled and the overall shape of the body is maintained between strict limits. One of the main ways in which this is accomplished is by the very careful monitoring of all the cells which make up each organ in the body and controlling their ability to divide and proliferate. There are many interacting systems which influence whether a cell remains in the resting phase or enters the cell cycle. In addition, there are other mechanisms that can be activated which induce the cells to commit suicide and destroy themselves. In fact, many of these mechanisms are still poorly understood. As there are billions of cells in the human body, the fact that cancer usually occurs in middle to old age demonstrates just how effective are the control mechanisms in the body.

However, should one of those cells become mutated and no longer respond to the controlling mechanisms, then it may begin to grow, divide and proliferate more than it should. As time goes by it may accumulate further mutations which make it even less susceptible to the body's growth controls. If it is not checked, then this abnormal cell will soon become dozens, then thousands and eventually millions of abnormal cells. It has been estimated that a tumour which is 0.5cm in diameter may contain 500 million cells. It is at that stage that the sheer mass of abnormal cells becomes noticeable and this is called a cancer or, in the case of an abnormal bone marrow cell, a leukaemia.

Chapter 3
WHAT IS CANCER?

What Distinguishes a Cancer Cell From a Normal Cell?

In essence a cancer cell behaves differently to its normal equivalent. The main feature that distinguishes the two is the uncontrolled proliferation of the cancer cell compared to that of a normal cell. This does not necessarily mean that the cancer cell grows and divides any faster than a normal cell, but rather that it does so more often. Many normal cells spend most of their life in the body in a resting state, locked into the G1 phase of the cell cycle. This is often referred to as being in the Go phase. Tumour cells also exhibit another aberrant behaviour in that they are able to spread around the body, settle in other organs and then form secondary tumours; most normal cells of course do not show this type of behaviour. This spread of tumour cells is called **metastasis** and poses the greatest difficulty in treating cancer. After all, if a tumour arises in an organ and does not metastasise, then it should be curable simply by surgery (cutting it out) or radiotherapy (destroying it with high dose radiation). These options are very much more difficult to apply successfully when the cancer cells have spread to other parts of the body and formed secondary tumours.

This overview of cancer, however, disguises the fact that there are several different processes which all contribute to the uncontrolled proliferation and eventual metastasis of tumour cells. To understand what really defines a cancer cell it is important to consider each of these underlying processes. The main ones discussed here are inhibition of **programmed cell death**, altered **proliferation** controls, blocks in **differentiation**, **immortalisation**, **angiogenesis** and **metastatic spread**. It should be of little surprise to learn that aberrant changes in each of these processes is dependent on abnormal protein manufacture, and this in turn is due to mutations of a number of genes. The conversion of a normal cell into a cancer cell is often termed **malignant transformation**.

Cancer is a Genetic Based Disease

It is now well established that cancer is a genetic based disease. That is to say, it is a disease which involves the genes. This is even true of situations where an external agent such as a virus is involved, because the action of the **virus** has now been shown to interfere with the normal activity of the genes in the cell. There are several lines of evidence that point to genetic information being involved in malignant transformation of cells.

One of the first indications that causative factors in cancer could involve the genes came from observations of clusters of cancer in family groups. It is clear that cancer does not occur randomly in families, but that some family groups appear to have less cancer than average and some have a very much higher incidence than average, particularly at younger ages. This familial relationship is very clear in some rare types of cancer, such as **retinoblastoma**. Retinoblastoma is an unusual tumour of the eye which occurs primarily in children. Some cases appear to occur randomly in the population but are rare and affect only one eye. However, other cases appear to occur in particular families. These familial forms of the disease often result in tumours forming in both eyes and there is a very high incidence of the cancer within the affected family group. Studies of this tumour have indicated that a genetic change is involved and that it can be inherited from one generation to the next. In this case it was demonstrated that the gene in question was situated on chromosome 13. This was subsequently isolated and called the *RB1* **gene**. As mentioned in Chapter 2, chromosomes occur in pairs and it was shown that the *RB1* gene was either missing or mutated on one chromosome 13. It was the inheritance of this defective gene that caused the problems. The normal copy of the *RB1* gene on the other chromosome 13 was able to maintain normal behaviour in the cells of an affected individual but, if as a result of a random mutation this was also inactivated, then that cell would begin to exhibit abnormal behaviour. The chances of such a mutation occurring in the *RB1* gene would be millions to one, but there are of course millions of cells in the body. This meant that in a cell where one copy of the gene had already been inactivated, the chances of a further mutation in the normal gene producing cancer was very high, maybe 1 in 40. On the other hand, in unaffected individuals the chances of acquiring mutations which inactivated both genes in any one cell were very remote and the incidence of these sporadic tumours is probably only 1 in 30,000.

The mutations that occur in genes do not all have the same potency in producing cancer. The *RB1* gene codes for an important protein in the cell which has regulatory functions that control the activity of other proteins and the activation of many other genes. Its effect on the cell is therefore amplified and if this is missing from the cell due to mutation/inactivation of the two copies of its gene, then it can have a very profound effect on the entire activity of the cell.

Other genes that can be mutated and inherited through the generations but which have a more subtle effect in cancer development have also been identified. An example of such a gene is *APC* which can predispose individuals carrying the mutated gene to develop a form of colon cancer. The *APC* gene has been located on chromosome 5 and it appears that when only one copy of the gene is mutated it can have an effect on the epithelial cells lining the colon. This is because the mutation in the gene results in the production of a slightly altered protein which has abnormal activity. This is a subtle effect and many affected people can reach adulthood and pass on the altered gene to their children before its effect manifests itself as a life threatening cancer. The first effect of the activity of this gene is to cause the formation of hundreds of polyps in the colon. These are **precancerous growths**, which by themselves are not dangerous but they often, as a result of mutations in other genes, evolve malignant cells which subsequently develop into tumours.

The inheritance of factors which predispose individuals to develop cancer is therefore good evidence that there is a genetic basis to malignant disease. This conclusion has also been supported by the realisation that tumours are generally **monoclonal**. What this means is that a single cell in the body, after it acquires sufficient mutations to become malignant, will proliferate to such an extent that all of its progeny eventually form a tumour mass. The information that makes the cell malignant is therefore passed on to all of the cells that have arisen from that original single cell. This has been demonstrated by the identification of characteristics which are common to all of the malignant cells in the tumour. An example of such a characteristic is the presence of particular forms of enzyme. An enzyme such as glucose-6-phosphate dehydrogenase can exist in one of two different forms. The gene coding for this enzyme is located to the X chromosome and some women can therefore have genes for both forms of the enzyme, as they have two X chromosomes. In such individuals half of their cells produce form A of the enzyme and the other half produce form B. This is a purely random event and has no relevance in itself to cancer. However, this is a useful marker for studying cancer cells. If a tumour develops in such a woman, then studies of which form of glucose-6-phosphate dehydrogenase is present in the malignant cells have shown that all of the cells always produce the same form of the enzyme. If the tumour cells had arisen individually then some should contain form A and some form B, but they do not. The cells have all arisen from a single cell and therefore they all synthesise the form of the enzyme that was present in the original parent cell.

Another example of a marker which demonstrates the clonal relationship of all of the cells that make up a tumour is the presence of an abnormal chromosome. Mutagenic agents often cause alterations to the structure of chromosomes and hence rearrangements of the DNA that compose those chromosomes. These abnormal chromosomes are found only in the malignant cells in the body and not in the normal cells. Although chromosomal abnormalities increase in number and complexity as a tumour evolves it is often possible to identify one or more abnormalities that occurred early in the development of the tumour and these are found in all the malignant cells. An example of such an abnormal chromosome is the **Philadelphia chromosome** which was first identified in the white blood cells of patients with **chronic myeloid leukaemia**. This chromosome is a hybrid formed by the joining together of pieces of chromosomes 9 and 22. This abnormal chromosome is observed in all of the malignant cells of such patients, but not in their normal cells. This type of leukaemia is characterised by a chronic phase, lasting several years, which is controlled by chemotherapy but eventually evolves into a very aggressive malignancy, known as a blast crisis, which is often fatal. However, the Philadelphia chromosome is observed in all of the leukaemia cells in both the chronic phase and during blast crisis. This again indicates that all of the cells that comprise a tumour or leukaemia originated from a single cell and the information that makes them behave abnormally has been inherited, which means it is genetic.

One of the most characteristic features of cancer cells is the presence of abnormal chromosomes. One of the features that is used to diagnose the presence of cancer cells in a cervical smear is the occurrence of abnormal chromosomes in the cells. This in itself is good evidence that cancer is a genetic based disease, because alterations to the structure of the chromosomes must have altered the DNA and this will result in changes to at least some of the genes. There are many types of **chromosome abnormality** that can occur in a cancer cell. These are usually classed as numerical or structural abnormalities. Some examples of chromosomal abnormalities that have often been observed in different types of cancer are listed in Table 3.1. It can be noted from this brief listing that different chromosome abnormalities often occur in different types of cancer, though sometimes the same abnormalities do occur in different cancer types. The frequency with which the different abnormalities occur also varies considerably between the abnormalities themselves and between the different tumour types.

Table 3.1 Some chromosome abnormalities observed in different cancers.

Cancer type	Chromosomal abnormality	Incidence
Breast carcinoma	Structural alterations to chromosome 1	80% of cases
Ewings sarcoma	Translocation between chromosomes 11 and 22	90% of cases
Kidney carcinoma	Deletions or translocations involving chromosome 3	80% of cases
Melanoma	Deletions or translocations involving chromosome 1	60% of cases
Ovarian carcinoma	Structural alterations to chromosome 1	80% of cases
Retinoblastoma	Deletion of region q13 of chromosome 13	20% of cases
Small cell lung carcinoma	Deletion of part of chromosome 3	90% of cases
Chronic myeloid leukaemia	Translocation between chromosomes 9 and 22	All cases
Burkitt lymphoma	Various translocations involving chromosome 8 region q24	All cases
Diffuse large cell lymphoma	Various translocations involving chromosome 8 region q24	20% of cases
	Translocation between chromosomes 14 and 18	40% of cases
Follicular centre lymphoma	Translocation between chromosomes 14 and 18	60% of cases

Numerical chromosomal abnormalities may be observed due to the duplication of the complete set of chromosomes. This can produce cells which instead of containing 46 chromosomes, may have 69 or 92 chromosomes. Alternatively, individual chromosomes may become duplicated so that the cell contains extra sets of all of the genes that the duplicated chromosomes carry. This could mean that if there were three instead of two chromosomes of a particular type, then the cell would have three not two copies of all of the genes on those chromosomes, which could in turn lead to the synthesis of 50% more of the proteins coded for by those genes. Alternatively chromosomes could be lost from the cell. This has been observed in patients who develop retinoblastoma where one copy of chromosome 13 is sometimes missing from their malignant cells.

Amongst the structural abnormalities that might occur are **translocations** where two or more chromosomes swap pieces with one another and thus produce hybrid chromosomes. An example of this was mentioned above where in chronic myeloid leukaemia cells, chromosomes 9 and 22 swap pieces to produce two hybrid chromosomes, of which the most abnormal has been called the Philadelphia chromosome. These abnormal chromosomes are designated as t(22;9) and t(9;22) to indicate they have been formed by a translocation event and to indicate their origins as being from chromosomes 9 and 22. This results in the formation of a fusion gene *BCR/ABL*, by joining together parts of two genes, one from chromosome 9 and one from chromosome 22. This results in the synthesis of an abnormal bcr-abl **fusion protein**, which has unusual activity within the cell. Other abnormalities include the **deletion** and **amplification** of small regions of a chromosome. Figure 3.1 shows a karyotype with a very abnormal set of chromosomes, present in a cell from a patient with malignant **lymphoma**. In this case, there are many more chromosomes than the normal 46. There are three copies of chromosomes 3, 5, 6, 7, 18, 20 and an extra X chromosome. This patient is a male as can be seen by the presence of a Y chromosome. A number of structural abnormalities are also apparent, several of which are indicated by arrowheads in Figure 3.1. There are also two **marker chromosomes** present which are so abnormal that their origins are uncertain.

Cells from a different lymphoma would probably contain a different set of abnormalities, but there are likely to be some which are common such as the t(14;18) translocation (Table 3.1). The main point is that when malignant cells are examined they almost always contain altered chromosomes, either different numbers of chromosomes or structurally altered chromosomes, or both.

Another body of evidence that has supported the idea of cancer being a genetic disease is concerned with the effects of cancer causing agents **(carcinogens)** on DNA. It has been demonstrated that many chemical and physical carcinogens that cause cancer also damage DNA and that their potency in causing cancer is often related to the amount of damage they can cause. These mutagenic effects are sometimes due to a direct interaction between the carcinogen and the DNA, but can also be due to indirect effects. For example, many chemical carcinogens are only converted into a dangerous form once they have entered the cell and have been altered by the cell's own enzymes.

Together these lines of evidence support the idea that cancer is a genetic based disease and that mutation of the DNA that makes up the genes results in abnormal protein synthesis, which in turn alters the behaviour of the cell. We therefore need to consider what are the key changes in the cells behaviour and to discuss how mutations can bring them about.

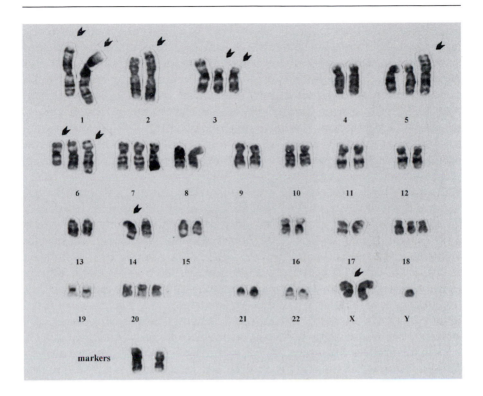

Figure 3.1 Karyotype of chromosomes from a malignant lymphoma cell. There are many extra chromosomes visible as well as structural alterations to several of the chromosomes (indicated by arrows). Two marker chromosomes are also present but are so altered that it is not possible to identify their origins.

Programmed Cell Death — The Key to the Formation of a Cancer Cell

Some years ago an interesting phenomenon was observed, which at first did not seem to most people to be relevant to cancer. However, it is now recognised as being of prime importance in the evolution of a malignant cell. This process is called programmed cell death, or **apoptosis**. There are distinct processes by which cells can die. They can be killed and die by necrosis. Alternatively cells can commit suicide and undergo apoptosis. The role of apoptosis in normal cell physiology, as well as in pathological conditions, has become an important topic of research during the past few years.

It is widely accepted that there is a homeostatic control of cell numbers in many tissues, which is the result of a balance between cell proliferation and apoptosis. This process of programmed cell death is associated with a series of morphologically distinct events. In normal tissues this appears to occur in only a few cells in the population, but because it is a very rapid event when it does occur, it is difficult to be certain just how common it may be. The role of apoptosis in such situations may be seen as a way of maintaining the optimal numbers of cells in a population, as in maintaining the number of white blood cells in the body. Alternatively it may function during embryonic development to allow for groups of cells to be removed during the moulding of the shapes of body parts.

Apoptosis is a very ordered process, unlike necrosis where the cell randomly disintegrates. During apoptosis the cell begins to condense. In the nucleus there is a condensation of the chromosomes and a gradual fragmentation of the DNA. Eventually the entire cell fragments into small membrane bound bodies which are engulfed by scavenger white blood cells and completely degraded. In this way all of the cellular components are disposed of in a controlled fashion and then recycled.

Of particular interest is the finding that apoptosis also plays a defensive role within the body. When a cell becomes damaged, particularly when its DNA is mutated, then it appears to have one of two choices. Either it repairs the damage or, if this is not possible, it commits suicide and destroys itself through apoptosis. This process clearly has important implications for the development of cancer. If mutations occur in genes that are associated with malignant transformation and if the cell is unable to repair them, then the cell would commit suicide. In this way cancer should never occur, but it does. The reason why cancer exists appears to be because on rare occasions mutations occur in genes that code for proteins which are involved in regulating the apoptosis process. These mutations greatly reduce the cells' ability to commit suicide. Although this does not produce a cell which is malignant, it does produce one that can now acquire further mutations and survive. The first step in creating a cancer cell therefore appears to be a blockage of the apoptosis process.

A number of genes have been identified that code for proteins which are important in controlling apoptosis and some of these have been found to be mutated in cancer cells. In malignant lymphoma, one of the most common mutations affects the *BCL-2* gene. This gene is normally switched off in white blood cells, thus making them susceptible to apoptosis. However, in lymphoma cells a chromosome translocation involving chromosomes 14 and 18 causes the *BCL-2* gene on chromosome 18 to be switched on indefinitely. The resulting inappropriate synthesis of the bcl-2 protein is sufficient to block the cells from entering apoptosis and thus allows them to survive the acquisition of further mutations. The inappropriate synthesis of the bcl-2 protein does not however result in a malignant cell and the use of very sensitive molecular biology techniques has demonstrated that mutations of this gene can occur in normal cells in some individuals who are free from cancer.

A block in apoptosis is therefore an essential first step in creating a malignant cell, but it does not in itself produce a malignant transformation of the cell. Malignant transformation requires the mutation of other genes which directly affect aspects of cell behaviour such as proliferation and metastasis.

Cell Proliferation

The mechanisms that control progress of the cell through the cell cycle have been the subject of an enormous research effort during the past two decades. This phenomenon is central to all life, in that all cells have to grow and then divide to proliferate. It now appears that there are striking similarities between the controlling mechanisms used by a wide spectrum of life. Although animal cells are certainly more complex than simple yeasts, it is clear that when cell cycle progress is considered they share a number of basic principles.

Details of the proteins that control progress through each phase of the cell cycle are beyond the scope of this book, but there are some interesting links between these proteins and cancer. It is now clear that many of the proteins that are altered in cancer cells, because of mutations in their genes, are proteins which are involved in regulating the cell cycle.

Cells are usually stimulated to grow by the action of **growth factors**. These **hormone** messengers generally circulate in the environment outside the cell and will trigger a growth response by the cell if they bind to a **receptor** protein on the surface of the plasma membrane of the cell. The binding of a growth factor to a receptor will then activate a train of events involving a series of **signalling proteins** (and maybe some smaller molecules inside the cell). This cascade of proteins can be visualised as a series of on/off switches, which are all gradually switched on and eventually take the growth signal into the nucleus. In the nucleus there are other proteins called **transcription factors** which, when activated, can bind to DNA and switch various genes on or off. There are other complications to this basic system, in that there are several different signalling pathways in any cell, some of which are inhibitory not stimulatory, and they often influence one another. However, this simple version is sufficient to understand how mutations could affect the regulation of cell proliferation.

It has been found that certain genes are associated with the emergence of cancer in animals and these have been called **oncogenes**. Two of the these oncogenes have now been demonstrated to code for growth factors. The *SIS* gene codes for platelet derived growth factor and the *FMS* gene codes for colony stimulating factor 1 growth factor. A mutation which resulted in overproduction of these proteins could clearly play a role in stimulating cell proliferation, because the more growth factor available to the cell and to its neighbours, the more likely they would be stimulated to grow and divide.

Other oncogenes have been found to code for receptor proteins that bind the growth factors. An example of one of these genes is *ERB-B2*. It has been found that as a result of a mutation the number of copies of this gene can be so increased that a cancer cell may contain hundreds of copies not just two. This in turn results in the gross overproduction of its protein product. The fact that these tumour cells are covered in these receptor proteins presumably means that they are very sensitive to even low levels of growth factors and therefore are readily stimulated to grow and divide. The amplification of the *ERB-B2* gene is a common mutation in breast cancer, particularly in those tumours that have a poor prognosis.

The activation of a receptor by a growth factor leads to the growth signal being passed through a chain of signalling proteins inside the cell. Again many of the oncogenes have been found to code for signalling proteins. One of the most common mutations in human carcinomas involves the *RAS* family of genes. These are three related genes, each of which codes for a signalling protein. A mutation of a single nucleotide base in one of these genes can result in the substitution of a different amino acid at just one point in the ras protein. This has the effect of altering its activity, so that instead of acting as an on/off switch, it is converted into a switch that is permanently switched on. This means that even in the absence of growth factors, it will initiate a growth signal within the cell.

Eventually the growth stimulation signal will reach the nucleus where transcription factors can be activated to stimulate or inhibit the activity of other genes. Not surprisingly some of the oncogenes have also been found to code for transcription factors. A case in point is that of **Burkitt's lymphoma**, an aggressive tumour which is often observed in children in Africa. In these lymphoma cells a chromosome translocation results in the activation of the *MYC* gene. This in turn causes the myc protein to be synthesised throughout the cell cycle, which is unlike the situation in normal cells where it is produced only during a short period of the cell cycle. This protein has a powerful effect on the activity of a number of genes and by prolonging its influence it encourages the cell to continually grow and divide.

These studies indicate how mutations in genes that code for proteins which have a function in controlling the movement of the cell through the cell cycle can affect the behaviour of the cell. Clearly cells that no longer need to have a stimulus to encourage them to grow and divide will do so without control. Eventually this uncontrolled proliferation will produce a colony of cells that will result in a tumour.

Immature Cells and Differentiation Blocks

As mentioned in Chapter 2, blood cells develop in the bone marrow where they gradually take on the final characteristics that define them as red blood cells or white blood cells. This process is called differentiation and occurs in many cell types in the body, not just bone marrow cells. Most differentiation occurs during embryonic development when the body is being formed, but it is also an important process in the adult body where cells have to continually replace those that have been lost. Blood cells and epithelial cells of the skin are two good examples of the latter. The undifferentiated cells which give rise to these various differentiated types of cells are called stem cells. The stem cells have the capacity to proliferate extensively but do not show the specialised characteristics that define their fully differentiated progeny. Cells may undergo several rounds of growth and division as they gradually take on their final differentiated form. In other words, until the cells become fully differentiated, which is often associated with a loss of proliferative capacity, the cells can still respond to growth factor signals to grow and divide.

It is interesting to note that many tumours are composed of malignant cells which appear to be less differentiated than the normal cells found in the tissue from which the tumour cells arose. It is often unclear whether the tumour cells have arisen in part because they have been blocked at an early differentiation stage or whether they have reversed some of the differentiation characteristics by becoming **dedifferentiated**. Nevertheless the degree of differentiation exhibited by tumour cells is a useful prognostic indicator in some tumours and is used in the classification schemes for many tumour types. In breast cancers for example, tumours that are composed of more undifferentiated types of cell are regarded as having a worse outlook than the more differentiated tumours. In this situation perhaps 20% of women will survive the

41

former group of tumours, whereas 80% of women may survive the latter.

An example of a malignancy which is characterised by cells which appear to have been blocked at an early stage of differentiation is **acute promyelocytic leukaemia**. This is characterised by cells in the blood that appear to be an immature myeloid type, which usually do not occur outside of the bone marrow. It appears that as a result of a block in their differentiation, these cells continue to proliferate in the bone marrow until their numbers are so great that they spill out into the blood, thus forming a leukaemia. These cells all have a characteristic chromosome abnormality which is a translocation involving chromosomes 15 and 17. It has since been demonstrated that this rearranges a gene called *ERB-A*, which in turn blocks the differentiation programme in these cells, causing them to remain as immature promyelocytes.

The fact that many tumour cells are apparently immature has a number of implications both for diagnosis and also for treatment. One interesting possibility is that if such a differentiation block could be released, then the cells might continue their differentiation programme which would result in non-proliferating cells. This possibility has been investigated and drugs related to dimethylsulphoxide, which will release this differentiation block, have been found to have promising effects in patients with acute promyelocytic leukaemia.

Immortalisation

One of the interesting characteristics of cancer cells is that unlike normal cells they do not appear to age. It is unclear at present why tumour cells do not show signs of ageing, but it is known that if normal cells are grown in culture they can be immortalised by adding extra genes to them. This technique, which is called **transfection**, enables one or more additional genes to be inserted into the DNA of the cell. Interestingly the genes which appear to have this ability are all classed as oncogenes and include the *MYC* gene, which has been implicated in the development of malignant lymphoma.

We therefore have further evidence of the complexities of malignant transformation. Here we have a gene, the *MYC* gene, which has the capacity to immortalise a cell, but has also been implicated in cell cycle regulation. How these abilities are interrelated is still unclear.

Angiogenesis — Feeding the Tumour

Two of the problems that cancer cells face when they proliferate and start to form a tumour mass, are the requirement for a supply of nutrients and oxygen, and the disposal of waste chemicals and carbon dioxide. Tissues are serviced by the blood supply which is adequate for the existing cell population. However, when a colony of malignant cells emerge, then their demands outstrip the available resources. This is particularly so in the centre of the growing tumour and without an improved blood supply these cancer cells will begin to die.

If the cancer cells are to be successful in producing a tumour, they therefore need to encourage the growth of blood vessels into the tumour mass. This stimulation of new blood vessel formation is called **angiogenesis**. Generally normal cells would not stimulate the growth of blood vessels, but mutations in the cancer cells may result in the synthesis of proteins which stimulate this process. Examples of such proteins are types of fibroblast growth factor which have a stimulatory effect on the endothelial cells that make up the blood vessels, as well as on fibroblasts.

The **vascularisation** of the tumour mass thus ensures that the growing demand for nutrients and oxygen, from the increasing numbers of cancer cells, is satisfied. This process has one other very important effect. The increasing network of blood vessels within the tumour mass means that many cancer cells are brought into very close proximity with the blood supply thereby facilitating the movement of cancer cells from the tissue in which they arise and into the blood or lymphatic circulation. This is of clinical importance, because it is the first step in the spread of the cancer to other parts of the body.

Metastasis — the Spread of Cancer

Perhaps the major factor that frequently results in cancer being a fatal disease is the spread of the cancer cells from their site of origin to other parts of the body. The emergence of multiple **secondary tumours** in different organs can pose serious difficulties to successfully treating the disease. It is important to remember that cells do not usually move around the body. There are very strong forces ensuring that they remain where they are supposed to be; liver cells stay in the liver and do not migrate to the heart or lungs. It is the ability of cancer cells to move to other tissues which is perhaps one of their most abnormal characteristics.

The process by which tumour cells detach from the primary tumour mass, enter the blood or **lymphatic circulation** and eventually settle down in another tissue to form a secondary tumour has been studied extensively in recent years. It is a complex phenomenon which is incompletely understood, but involves overproduction of certain proteins and underproduction of others. For example, the gene *NM23* has been shown to be inactivated and therefore its protein not synthesised in a number of tumour cells which are able to metastasise. Many of the genes that are activated in such cells produce proteins which act as enzymes that degrade the connective tissue which surrounds the cells. Genes that are inactivated appear to code for proteins which are located in the plasma membrane of the cells and act as linkers that hold adjacent cells together. The overall effect of these changes is to reduce the adhesion of the cancer cells within the tissue, and thus increase their potential mobility.

Overview of Cancer Cell Evolution

One of the most interesting facts about tumours or leukaemias is that they arise from a single mutated cell. So out of the billions of cells that make up a liver, the lining of the gut, or a bone marrow, it only needs a single cell to acquire sufficient genetic alterations to enable it to proliferate out of control and gradually develop into a tumour. It is a sobering thought that the large tumour masses that occur in some patients, which contain billions of cells, have arisen from just a single mutated cell. This in part explains why it often takes so long for a tumour to become apparent.

However, the main reason why cancer takes so long to develop, and is usually a disease of old age, is that so many mutations have to occur before the cell is truly malignant. Furthermore, these mutations need to occur in the right order so that the cell can survive during the early stages of malignant transformation. As these mutations are rare events, it can take a considerable time before a cell has acquired the required number to affect its behaviour. These events are summarised in Figure 3.2 and clearly demonstrate the importance of blocks in apoptosis that allow malignant transformation to occur.

It should be noted that the prevention of apoptosis may have another important effect. There is now evidence that a block in apoptosis is a crucial event underlying the emergence of drug resistance in cancer cells. This will be discussed in more detail in Chapter 6, but the importance of drug resistance in the treatment of cancer patients cannot be overstated. For all of the complexities of the cancer cell, there are currently some very effective drugs available which would certainly kill them but for drug resistance.

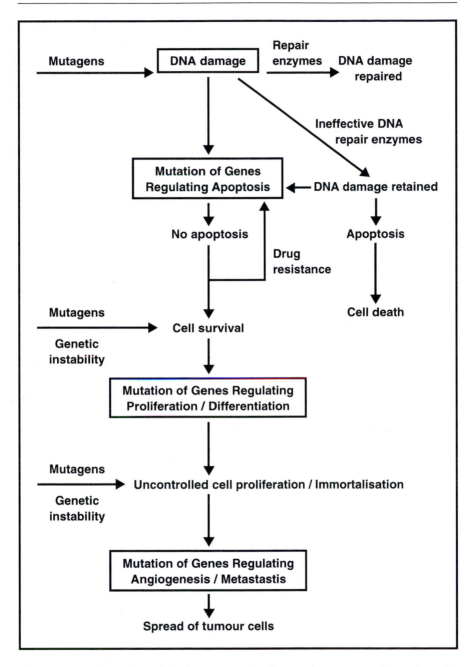

Figure 3.2 Overview of the key events in the malignant transformation of a cell into a cancer cell.

Chapter 4
CANCER TYPES

Introduction

Not all cancers are the same. In some ways it is very misleading to talk about cancer being one disease. It is in fact a whole series of diseases which often share a few general symptoms, mainly to do with the poorly controlled proliferation of cells. Similarly in previous centuries fever was thought to be a single disease but with the advent of microbiology it soon became apparent that was a common symptom of many different diseases, each with a different cause and often requiring different treatments. So it is with malignant disease. There are many different causes of cancer, affecting different cells in the body, which may be treated in different ways with different outcomes.

It is important to remember that similar types of tumour may occur in different patients and yet have quite different outcomes. Figure 4.1 shows an example of this situation. In these patients swellings were noticed in their necks. In both cases these were caused by lymphoma, a proliferation of white blood cells in their lymph nodes, which caused the nodes to increase in size. The patient in Figure 4.1A was, however, diagnosed with a form of lymphoma called Hodgkin's disease, which usually responds very well to treatment and the patients are often cured. Whereas the patient in Figure 4.1B was diagnosed with non-Hodgkin's lymphoma. This type of lymphoma does not always respond to treatment and the majority of patients with this malignancy eventually die from their disease.

Tumour Progression

Tumours do not develop overnight. It takes many genetic alterations to produce a fully malignant metastasising cancer cell. There is a gradual change or progression in the form and behaviour of these cells and the speed with which changes occur varies considerably from cell to cell. Generally such changes lead to the growth of a more malignant tumour, and only rarely do the tumour cells become less active and the cancer disappear.

The fact that tumour cells may acquire further mutations can lead

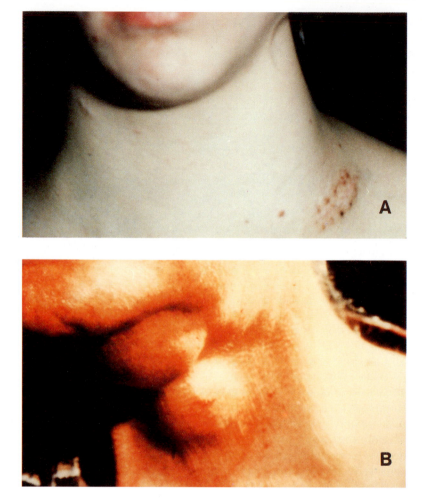

Figure 4.1 Patients with swollen necks due to the presence of malignant lymphomas. These were diagnosed as (A) Hodgkin's disease, (B) non-Hodgkin's lymphoma.

to mixed populations of malignant cells in the tumour. It is difficult to grade a tumour if part of it contains malignant cells which still exhibit characteristics of their original parent cells and others that have lost many of those characteristics. The latter appear to be more immature and are probably more aggressively malignant. This heterogeneous nature that many tumours exhibit can present problems for the pathologist when making a diagnosis. As mentioned in Chapter 3, the more differentiated the tumour cell, the more likely it is to respond to therapy. It is therefore important that the oncologist has a good idea of the tumour grade, as this could indicate how well the treatments will work.

When viewed down a microscope, significant changes in the organisation of a tissue can indicate the presence of a tumour. Some examples of these are shown in Figure 4.2. Figure 4.2A is a histological section which shows the well ordered appearance of the surface of a normal colon. When epithelial cells become malignant in the lining of the colon their proliferation can result in an increase in cell numbers and disruption of this ordered structure (Figure 4.2B). A similar situation can occur in the breast. Here a number of ducts run through the breast tissue and each is lined by a layer of epithelial cells. In the normal breast this lining is only two cells in thickness (indicated by arrowheads in Figure 4.2C). The most common forms of breast cancer involve these epithelial cells and result in their proliferation so that the linings of the ducts in the breast become many cells thick (indicated by arrowheads in Figure 4.2D). This is also often associated with these duct-like structures spreading into and destroying adjacent tissue.

There are several stages that a cell must go through to become a fully malignant cancer cell. This is mirrored in the behaviour of the tumour itself. For example, the earliest stage of colon carcinoma development appears to involve unusually high levels of proliferation of epithelial cells. These are not malignant cells, but appear to be a premalignant state which is often associated with mutation of one of the copies of the *APC* gene. This may then develop into a benign adenoma tumour if the second copy of the *APC* gene is lost or mutated. However, further mutations in *RAS*, *DCC* and *P53* genes can result in progression until there is a fully malignant carcinoma. Further mutations and changes in the behaviour of the cells can then result in metastatic spread. These include invasion of the surrounding tissue and penetration into the small blood vessels and lymphatic channels. The cells then enter the circulation and, if they survive the experience, may settle in the capillary beds of distant tissues. They are then able to invade, colonise and form secondary tumours in those tissues. There are therefore several significant changes in the behaviour of these cells all of which are dependent on mutations in their genes.

Benign versus Malignant Tumours

Not all tumours are malignant. That is, not all tumours contain cells that proliferate out of control, invade surrounding tissue, spread to

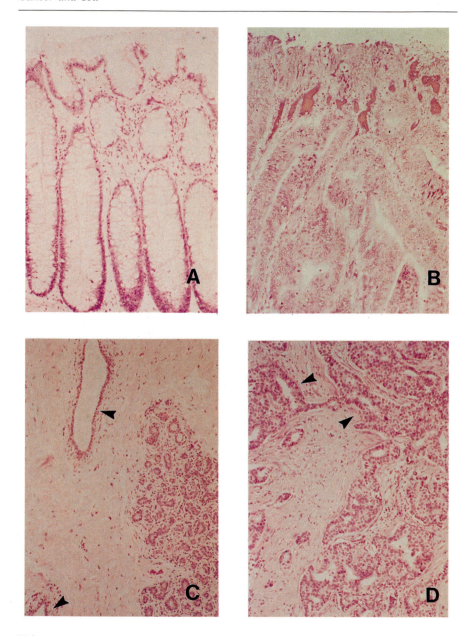

Figure 4.2 Histological sections through normal and tumour tissue as examined under the microscope. (A) normal human colon, (B) human moderately differentiated adenocarcinoma, (C) normal human breast, (D) human breast carcinoma. The arrowheads in (C) and (D) indicate the ducts that run through the breast tissue.

other parts of the body and are life threatening. Benign tumours can arise in most tissues as a result of a localised proliferation of cells, but they do not invade surrounding tissue or metastasise. They are usually separated from surrounding tissue by a capsule of connective tissue, or protrude from the surface like a wart. The cells that comprise the benign tumour usually resemble their normal counterparts. They are classified into a number of types, of which papillomas and adenomas are among the most common.

Generally benign tumours do not present a problem to the patient and can be removed by surgery. This might be necessary if, because of their growth, the mass of the tumour begins to put pressure on nearby tissue, particularly nerves. An example of a benign tumour producing a problem would be one that grows adjacent to the optic nerve and through its growth eventually damages that nerve. This could lead to blindness in the patient. Other problems might occur if the benign tumour arises from cells which are part of an endocrine gland. If these cells produce a hormone, then a build up in their numbers and failure of normal control of secretion could result in an overproduction of that hormone. An increased level of a hormone in the blood could cause a number of problems to the patient.

Malignant cells can be distinguished from those in benign tumours by the presence of abnormalities, particularly abnormal chromosomes, and their ability to invade surrounding tissue. The malignant tumours are not surrounded by a well defined capsule and their cells tend to proliferate in a more disorganised manner to that seen in benign tumours. It should be noted however that benign tumours may subsequently develop into malignant cancers.

Carcinomas

As is discussed in Section II, many of the causes of cancer are environmental and therefore it is hardly surprising that most forms of malignant disease originate in surface lining cells. What many people fail to realise is that the layer of skin that covers the outside of the body is also continued inside the body. The epithelial cells which make up the skin, also line the throat, the stomach, the intestines, the lungs and the ducts running through many organs such as the breasts. These are the first cells to experience the insults of carcinogens in food, drink and air. They are the first to experience radiation in sunlight and are the first to be attacked by invading viruses. As the lining cells are constantly dividing to replace those lost, they are at particular risk. It is of little

Table 4.1 Types of tumours arising from cells of different tissues.

Tissue	Malignancy
Bladder	Carcinoma
Bone	Osteosarcoma
Bone marrow	Leukaemia, lymphoma, myeloma
Breast	Carcinoma
Cartilage	Chondrosarcoma
Colon	Carcinoma
Fat	Liposarcoma
Fibrous tissue	Fibrosarcoma
Lung	Carcinoma
Liver	Carcinoma
Muscle	Rhabdomyosarcoma
Nerve	Neuroblastoma
Ovary	Carcinoma
Pancreas	Carcinoma
Prostate	Carcinoma
Skin	Carcinoma, melanoma
Stomach	Carcinoma
Thyroid	Carcinoma

surprise therefore that the tumours arising from these epithelial cells comprise the largest group of all malignancies. The cancers that arise by malignant transformation of the epithelial cells are collectively known as carcinomas (Table 4.1).

There are sub-divisions within this group of cancers which reflect the differentiation state of the cells from which they arise. The squamous epithelial cells therefore give rise to squamous cell carcinomas, and basal epithelial cells give rise to basal cell carcinomas. The tumours may grow as flattened bodies or as wart-like outgrowths, in which case they would be classed as sessile or papillary carcinomas.

Classifications of this kind can be very useful, because experience has shown that carcinomas arising from one type of epithelial cell can respond differently to treatment. For example squamous cell carcinomas are more likely to respond to radiotherapy than basal cell carcinomas.

Leukaemias, Lymphomas and Myelomas

Most cells in the body are generally fixed in place within a tissue and when they become malignant they initially form a mass of cells locally, which is what we call a tumour. Only later will cells from the tumour mass become detached and float away in the blood or lymphatic system to spread to and colonise other tissues. Once there, the cells settle down to proliferate and form other tumour masses, the secondary tumours. However, some cells naturally travel around the body. These are the blood cells and they are formed in the bone marrow. When the bone marrow cells become malignant and begin to proliferate out of control they follow their natural behaviour and spill out into the blood. As many more of these malignant blood cells are produced than is normal, then the blood cell count can increase dramatically. This is what is known as a **leukaemia**, because of the increase in white blood cells (Table 4.1). The first thing that is obvious about these malignancies is that they mainly involve white and not red blood cells. As cancer is a genetic based disease, and as the red blood cells have lost their genetic material during their production they no longer have genes that can be altered. The only type of red blood cells that can form leukaemias are immature cells which have not differentiated far enough to lose their nucleus.

Leukaemias are classified into chronic and acute types. In the first case there is a gradual increase in white blood cells seen in the blood and the individual may initially be unaware of any significant change. The two main types of chronic leukaemia involve myeloid and lymphoid types of white blood cell and are called **chronic myeloid leukaemia** and **chronic lymphocytic leukaemia**. Whereas chronic myeloid disease is eventually lethal, patients with chronic lymphoid leukaemia can often live normal lives with the disease being controlled by oral chemotherapy. The **acute leukaemias** appear to be more aggressive forms of malignancy and there many different types. However, these forms of leukaemia often respond well to chemotherapy and half of all patients (who are often children) can be cured.

As part of the normal immune response, many of the white blood cells will settle down in lymph nodes to await any invading bacteria or viruses, which they will then attack. Malignant white blood cells which are located in the lymph nodes do not attack invading bacteria, but simply continue to proliferate and form what is called a soft tumour, or a lymphoma.

There are two broad categories of lymphoma. These are distinguished by the presence of an unusually large type of cell in the lymph node tissue. These giant **Reed-Sternberg cells** are characteristic of **Hodgkin's disease** and without them the lymphomas are classed as a **non-Hodgkin's lymphoma**, of which there are also many subtypes. Generally speaking Hodgkin's disease responds well to therapy and many patients are cured. This situation is different in patients with non-Hodgkin's lymphoma where the majority, though not all, die from their disease.

A white blood cell that is capable of secreting antibodies is called a **plasma cell.** Although this might be considered as a fully differentiated cell, malignant transformation of such cells produce a relatively aggressive disease. These cells tend to locate back to the bone marrow where the tumours they form are called **myelomas**.

Less Common Cancers

Cancers of the mesenchyme are called **sarcomas** (Table 4.1). The mesenchyme tissues include cells that make up or are found in bone, cartilage, fat deposits, smooth muscle and blood vessels. These tumours are often fast growing.

The pigment cells in the skin can also be transformed to produce a tumour called a **melanoma**. This tumour has become newsworthy in recent years because of its association with sunbathing in white individuals. It is also a much more aggressive tumour than the usual carcinomas of the skin.

Tumours of the nervous system are uncommon and occur mainly in early childhood These **neuroblastomas** arise in nerve cells that are dividing, but some do occur after birth in the specialised nerve cell layer in the eye and are called retinoblastomas. Brain tumours that arise in the adult usually originate in cells other than nerve cells, as nerve cells no longer divide in adult life. Astrocytes for example still have a capacity for proliferation and can form **astrocytomas**. Interestingly, tumours arising in the brain do not usually spread outside of the brain or spinal cord.

Cancers of the testis are uncommon, but nevertheless are among the most frequently occurring tumours in men aged between 20 and 35. These arise from the germ cells (the cells that develop into sperm) and are classed as **seminomas** or **teratomas**. Women can also develop teratomas from the germ cells (the cells that develop into ova) in the ovary, and these also tend to occur in younger women.

Chapter 5
WHY DO YOU DEVELOP CANCER?

Introduction

Cancer is a disease primarily caused by mutation of certain genes within a cell. This in turn changes either the amount of the proteins that they code for in the cell, or the structure of those proteins. When certain viruses infect cells they can interact with and prevent some of these cellular proteins from working, which is equivalent to mutating the proteins' genes. To understand why cancer develops we therefore need to consider what causes DNA to be altered in a cell. These causes can be divided into two general categories: inheritable and environmental.

It is clear that some genes can exist in altered forms that do not cause cancer to develop at an early age, but which predispose the development of cancer later in life. This means that an individual who possesses one of these slightly defective genes can pass it on to their children before it has a damaging effect on the individual themselves. In this way, a predisposition to develop cancer can be passed through the generations.

However, it appears that by far the most important causes of cancer are external factors. Numerous influences from the environment have now been identified as having a causative effect on cancer development. These include certain chemicals, different types of radiation and viruses. Chemicals and radiation can have a direct effect of mutating DNA. Viruses can interfere with the function of the cell's proteins by introducing their own genes into a cell; though in some animals, viruses can mutate the animal's own genes to produce cancer causing genes.

Heredity

The heritability of cancer has been inferred by studies of the incidence of malignant disease in the relatives of patients with cancer. The contribution of this inherited component to cancer incidence is often not as great as the inherited contribution to heart disease or mental disease. This correlation is not as easy to demonstrate as one might think. A few studies have been carried on twins brought up in the same household

Table 5.1 Genes associated with inherited cancers.

Gene	Inherited cancer
APC	Familial adenomatous polyposis
P53	Li-Fraumeni cancer family syndrome
NF1	Neurofibromatosis type 1
NF2	Neurofibromatosis type 2
RB1	Retinoblastoma
VHL	Von Hipple-Lindau syndrome
WT1	Wilms tumour

and therefore subjected to similar environmental influences. These have indicated that identical twins show a more similar cancer incidence that non-identical twins, but these observations rely on small samples and are far from convincing. Generally the incidence of cancer in relatives of affected patients is two to four fold greater compared to that in the general population. However, when studies have been carried out to identify clusters of cancers in affected families some very impressive observations have been reported, which have indicated that genes predisposing individuals to the development of certain types of cancer do exist and can be passed down the generations.

There are two general mechanisms by which an inherited predisposition to cancer can manifest itself. The first is an effect on a particular type of cell, which is particularly sensitive to the effect of a certain gene. The second is a more general effect, where one gene can have an effect on most cell types. The first type of mechanism will tend to result in an increased incidence of one type of cancer, but the second mechanism will result in an increased incidence of cancer generally.

One of the best documented cancers which has a pronounced hereditary component is retinoblastoma, which was discussed in Chapter 3, but there are others (Table 5.1). There is, for example, a rare form of colon cancer called **familial adenomatous polyposis** (FAP). In this condition individuals who have a defective *APC* gene develop hundreds of polyps in their large intestine. These polyps are not in themselves necessarily cancerous, but it is exactly this type of benign tumour (adenoma) that may transform into a malignant cancer. When so many of these occur in one individual there is clearly a greatly increased risk that one or more of them will indeed give rise to a carcinoma.

An example of a gene having an effect on several tissues is shown in the condition **xeroderma pigmentosum**. This is a condition caused by a defective gene which codes for a protein which repairs damage to DNA. This type of DNA damage is primarily caused by sunlight, but can also

Table 5.2 Genetic mutations and associated cancers produced in mice after exposure to chemical carcinogens.

Carcinogen	Mutation	Cancer
Benzpyrene	K-ras codon 12 (G > T or A)	Lung carcinoma
Dimethylbenzanthracene	H-ras codon 61 (A > T)	Skin carcinoma
Ethyl carbamate	K-ras codon 61 (A > T or G)	Lung carcinoma
Hydroxyacetylaminofluorene	H-ras codon 61 (C > A)	Hepatoma
Methylnitrosourea	H-ras codon 12 (G > A)	Breast carcinoma
	K-ras codon 12 (G > A)	Lung carcinoma
Vinyl carbamate	H-ras codon 61 (A > T)	Hepatoma

occur by exposure to chemical carcinogens. Individuals with this condition often develop skin cancers after exposure to sunlight, but may also develop internal tumours after exposure to chemical carcinogens in their diet. Mutations of the *P53* gene have been found in many types of tumours and it is therefore not surprising that mutant forms of the *P53* gene, which are inherited in some families, result in a high incidence of different tumour types in those families. This type of predisposition is called **Li-Fraumeni cancer family syndrome** (Table 5.1), and over half the members of a family affected by it may develop cancer.

Environmental — Chemical

Many studies of cancer have now demonstrated links between cancer and exposure of the body to external influences. Other chronic diseases such as heart disease are also under the influence of external factors, but it is only in cancer that changes in DNA occur and this is fundamental to the emergence of malignant disease. This damage gradually accumulates over a period of time and is caused by physical, chemical and infectious agents from the environment. Therefore to understand the effects of these environmental factors we need to know how their components interact with DNA.

Some chemicals that are present in the work place, in tobacco smoke and in food are known **carcinogens** that interact with and damage DNA. Among the most frequently mutated genes in both human and animal cancers are the *RAS* genes, of which there are three types (*H-RAS*, *K-RAS* and *N-RAS*). It is of considerable interest therefore that experimentally induced cancers in animals, produced by exposure to chemical carcinogens, have been found to contain mutated forms of the *RAS* genes (Table 5.2). The implication is that the presence of chemical carcinogens can cause mutations in a reproducible manner, which in turn play an important role in the evolution of malignant disease. It is also likely that particular chemical carcinogens may produce particular types of mutation. In human tumours, the *K-RAS* mutations in lung cancer are often different to those found in colon tumours. This suggests that the different chemical carcinogens that are either found in tobacco smoke, or in the diet, produce different types of mutation. Differences in the pattern of mutation of the *P53* gene have also been noted in different tumour types.

In some cases the original chemical is not dangerous but is converted into a carcinogen by the metabolism of the body's own cells and only then will it interact with DNA to damage it. This metabolised version of a chemical has been termed an **ultimate carcinogen** by some researchers. Its activation usually involves enzymes such as the **P450 cytochrome** system that are located in the cell. The intention of these metabolic processes is to make the original chemical more water-soluble so that it can be better excreted. Unfortunately this process also makes the chemical more reactive. As a result of this conversion, these chemicals are better able to bind to large molecules in the cell such as DNA. These **Phase I enzymic mechanisms** can vary from one cell type to another and also differ in their efficiencies between species. As a result, the influence of any carcinogenic chemical can vary and lead to the emergence of a different spectrum of tumours, particularly in different animals.

There are other **Phase II enzymic reactions** which detoxify many chemicals and can inactive chemical carcinogens. The genes coding for these enzymes are often switched off and only become active after receiving a stimulatory signal. A number of chemicals in plant foods appear to act in a way that induces the activation of Phase II enzymes. This may well be an important reason why diets which are rich in fruit and vegetables appear to offer a protective effect against developing cancer. A great deal of experimental work has focused on these issues in recent years. It has been shown, for example, that the element selenium can both prevent the carcinogenic activation and can also increase the rate of detoxification of chemicals such as 2-acetylamino-fluorene. Selenium in the diet appears to have a protective action against cancer.

There are thousands of potentially carcinogenic chemicals in the environment which are capable of damaging DNA in all types of cells and which therefore may contribute to the evolution of a wide spectrum of tumours. How significant their impact may be depends on a complex set of interrelated factors including the levels of chemical that the cells in the body are exposed to, what other carcinogens are present and what protective chemicals may also be available to reduce or nullify their effects.

Aflatoxin

Aflatoxin is a food contaminant which is produced by mould growing on food in storage, particularly in the tropics. It has been implicated in the development of liver cancer and may be a co-factor with hepatitis B virus infection. It has been shown in experimental systems that it is capable of causing mutations in the DNA by converting the nucleotide base guanine into thymine. This appears to occur frequently in a particular region of the *P53* gene, which has the effect of changing the composition of the p53 protein by one amino acid. This mutant protein appears less able to carry out its function in regulating cell cycle progress and entry into apoptosis and is therefore an important factor in producing a malignant cell.

Oxygen Radicals

It has also been suggested that the body might generate its own cancer causing chemicals as part of its normal metabolism. This is because oxygen, which is essential for living processes, can be converted into a very reactive form called an **oxygen radical**. These oxygen radicals are a by-product of a number of chemical reactions and can potentially bind to and alter other molecules including DNA. It is not clear how important this mechanism is in causing human cancer, but components of the diet such as vitamin C are known to be effective in eradicating oxygen radicals. Vitamin C is also known to have a protective action against cancer.

Environmental — Radiation

Radiation can damage DNA directly and therefore exposure to **ultra-violet light** that is a component in sunlight, or to ionising radiation (**X-rays** or **radioactive decay** of chemicals) can clearly increase the risk of developing cancer. Ultra-violet light has been primarily linked to cancers of the skin, whereas the more powerful ionising radiations can cause cancers deep within the body. The latter have frequently been linked to leukaemias but the release of radioactive forms of iodine from Chernobyl a few years ago resulted in the take-up of this radiochemical by thyroid glands in the nearby population. This produced a greatly increased incidence of thyroid cancer. It has been estimated that 3% of all cancers are associated with radiation and most of these are skin cancers.

Chromosomal analysis of melanomas has indicated that chromosome 17 is usually unaffected. At first sight this might be taken to indicate that the *P53* gene is not involved with the evolution of these tumours (The *P53* gene is located on chromosome 17). However, molecular analyses have indicated that the protein product of this gene often appears to be abnormal in melanomas. This is an interesting observation because animal studies have shown that when *P53* is inactivated in melanocytes they showed a high predisposition to develop melanomas after exposure to low dose ultraviolet B radiation. In these experiments the dose of radiation was so low that it would not normally produce melanomas in normal mice. It is known the p53 protein protects the cell from damage to the DNA because of its roles in regulating DNA repair and entry in apoptosis. This experiment therefore suggests that if mutations have occurred in a cell resulting in the inactivation of the p53 protein, then the cell would become susceptible to more widespread DNA damage caused by ultra-violet B radiation in sunlight.

Environmental — Infectious organisms

Cancer causing viruses have been studied extensively in animals throughout the twentieth century. The involvement of viruses in causing human cancer does not appear to be as widespread as it is in animals, but epidemiological studies have implicated a few viruses as being important causative agents in certain types of human cancer. The main ones are the **hepatitis B** and **hepatitis C** viruses which have been linked to liver cancer, **human papilloma viruses** (HPV) which are linked to cervical carcinoma and **Epstein-Barr virus** (EBV) which has been linked to lymphoma and nasopharyngeal cancer. There are also some rare forms of leukaemia which appear to be caused, at least in part, by **human T-cell leukaemia virus** (HTLV). These studies have further suggested that viruses do not cause cancer by themselves in human cancers, but rather act as important co-factors with other agents to transform normal cells into malignant cells.

The above viruses all have a direct effect on the cells they infect and contribute to their malignant transformation. However, viruses can also have an indirect effect as has been shown by **human immunodeficiency virus** (HIV) in **AIDS** patients. It appears that some forms of cancer can be eradicated by the body's own immune system and therefore do not normally present a problem. However, the effect of the AIDS virus is to destroy this immune protection and therefore allow these otherwise rare types of tumour to emerge. These include an unusual form of lymphoma and **Karposi's sarcoma**.

Interestingly, infection by the bacterium **Heliobacter pylori** has recently been linked to the development of stomach cancer, though it is unclear at present how this bacterium may contribute to the malignant transformation of these cells.

Epstein-Barr Virus (EBV)

EBV is a very large virus and appears to contain over 90 genes. Two of these genes, EBV-determined nuclear antigen (EBNA)-2 and latent membrane protein (LMP)-1, appear to be involved in the immortalisation and subsequent malignant transformation of cells. LMP-1 protein can apparently increase the amount of a cellular protein called bcl-2. The bcl-2 protein has been implicated in the development of lymphoma and at high levels appears to prevent the cell from entering apoptosis. Under these circumstances the infected white blood cells show enhanced survival and are therefore able to accumulate more mutations which can contribute to their malignant transformation.

There is also some evidence that a third gene, EBNA-5, may also interact with the p53 and rb-1 proteins within the cell. Another EBV gene which may be of importance is the BCRF-1 gene which, from its structure, appears to be a growth factor that may stimulate the white blood cells to proliferate.

Amongst the mutations that appear to be important in lymphoma is a chromosome translocation involving chromosome 8 and others. The result of the translocation appears to be an activation of the *MYC* gene. This event is observed in all cases of Burkitt's lymphoma and to a lesser extent in other forms of high grade lymphoma.

Hepatitis B Virus (HBV)

HBV is a relatively small DNA virus. One of its genes, which has been called the X gene, appears to be required for its ability to aid the malignant transformation of cells. When transgenic mice which contained this gene were produced, they showed a high incidence of liver cancer. It appears that the protein product of the X gene can bind to the p53 protein and presumably prevent it from functioning. Certainly complexes formed between the X gene protein and p53 protein have been identified in liver cancer cells.

The fact that it may take many years between infection by HBV and the occurrence of liver cancer indicates that the virus alone is insufficient to produce fully malignant cells. Other co-factors or genetic lesions also appear to be necessary.

Hepatitis C is a retrovirus, which means that like HIV and HTLV its genes are composed of RNA not DNA. It is unclear at present how it is able to induce malignant transformation of human cells, or how widespread it is in liver cancer.

Human Papilloma Viruses (HPV)

There is a high risk of women developing cervical cancer if they have been infected with HPV and the virus has been found to be present in over 90% of cervical tumours. More than 70 types of HPV have been identified, though only 23 infect cells of the cervix. Six of these have been implicated in cervical cancer. The high risk ones being HPV types 16, 18, 31 and 45 and the low risk ones being HPV types 6 and 11. It is still unclear how many women develop cancer as a result of HPV infection.

HPVs are DNA viruses, though they are much smaller and less complex than EBV. They appear to contain 10 genes of which two, termed E6 and E7, appear to be important in malignant transformation of human cells. Certainly E6 and E7 RNAs have been observed in most cervical cancers. Studies have been carried out to judge the effect of isolated forms of these genes on cells. It appears that they have the ability to immortalise a variety of cell types. They can also cooperate with certain other genes to fully transform normal cells into malignant cells. It now appears that the protein products of the E6 and E7 genes bind to and inactivate the cell's p53 and rb1 proteins respectively. Interestingly, where cervical carcinoma arises without the involvement of HPV, the tumour cells appear to contain mutated *P53* genes. We therefore have two mechanisms by which the p53 protein can be inactivated and thus contribute to the malignant transformation of the cervical epithelial cells.

This mechanism does not appear to be sufficient by itself to produce a fully malignant cell for it has been demonstrated that when a cell infected with HPV is fused to a normal cell, the resulting hybrid cell behaves normally. In other words the normal cell is able to contribute some factor which can counteract the p53 inactivation in the HPV infected cell. This again indicates the complexity of changes that are required to produce a malignant cell. In this regard both cigarette smoking and co-infection with the herpes simplex virus type 2 have been implicated in acting with HPV to produce cervical cancer.

Human T-cell Leukaemia Virus (HTLV)

This is a RNA virus which contains only a few genes and has been implicated in acute T-cell leukaemia/lymphoma (ATLL). One of these genes, the *tax* gene, is thought to be an important factor in transforming human cells. Transgenic mice containing the *tax* gene develop sarcomas. The *tax* gene protein appears able to affect the activity of a range of cellular genes. These include *FOS*, *MYC* and *SIS*, which have all been implicated in the malignant transformation of human cells. It can also increase the production of growth factors such as interleukin-2 and decrease the activity of one of the DNA repair enzymes, B-polymerase, which could result in the accumulation of mutations in the cell.

Interestingly, the levels of tax protein are low or even absent in ATLL cells. This appears to suggest that whatever the contribution made by the tax protein to the malignant transformation of these cells, it is likely to occur at an early stage. It appears that by themselves HTLV gene products are insufficient to produce a full malignant transformation, and other factors must be involved.

Chapter 6
CANCER DIAGNOSIS

Introduction

A number of symptoms may occur which first alert a person to the fact that something is not as it should be with their body. These are discussed in more detail in Chapter 10. Once these have been reported to their doctor and a clinical history taken, it may well be necessary to investigate the problem in more detail. This further investigation is very important as it should enable the presence of a suspected cancer to be proven. This **diagnosis** of cancer is the first step in its treatment and will require specialised hospital facilities.

The types of test used are described below. Exactly which of these tests will be carried out will depend on the type of cancer being investigated. It may be that on occasion certain tests may have to be carried out again, to ensure the best diagnosis. Today it is generally the case that the oncologist will discuss the diagnosis with the patient and explain the various approaches for treatment. It is important that a patient has an awareness of diagnoses and subsequent treatment options so that their concerns are minimised. The less people know about a problem, the more they worry about it. In treating cancer a positive attitude from the patient is an important weapon in dealing with the disease.

Biopsies

A biopsy is a sample of a tumour that is removed for further examination under the microscope. Examples of the types of changes that occur in tissues and which are indicative of tumour formation are shown in Figure 4.2. It is often the case that a biopsy can be obtained using only local anaesthetic, for example, when a piece of skin is removed or when a bone marrow sample is extracted by a needle. In fact, biopsy samples may also be removed from internal organs such as the kidney, liver and lymph nodes by the use of needles. By making use of ultrasound, it is possible to pinpoint the position of the suspect tumour very precisely and therefore ensure that the needle tip samples cells from the tumour. This approach is less traumatic for the patient and is useful in confirming a positive diagnosis. However, because it involves few cells being sampled, a negative result may mean that a further sample has to be obtained by open surgery.

If a large tumour is present in a tissue then the entire area may be removed by the surgeon and sent to the pathology laboratory to be examined in more detail. In an operation where a large tumour is to be removed, or where lymph nodes in the neck or armpits are going to be removed, the procedure is carried out under general anaesthetic. Under these circumstances it is normal for the patient to stay in hospital.

Once a sample is obtained by the pathologist for analysis, it has to be processed for histological examination. This usually involves fixing the material in formalin, embedding it in paraffin wax and then cutting very thin slices (sections). These sections are then stained and examined under the microscope (Figure 4.2). This procedure may take two to three days and therefore a patient will need to wait a while for the result of their biopsy. In the case of bone marrow biopsies, there is an additional requirement to remove bone calcium before analysis which can result in a delay of up to one week.

In some cases the pathologist may decide that further analysis is required, and therefore more sections will be prepared from the biopsy material. These will then be subjected to different staining protocols to provide extra information. This of course will delay a report but could provide very valuable information about the tumour that has been biopsied. For example, if a lung tumour mass has been biopsied, it could reveal whether the tumour was a lung cancer, or whether it was a secondary tumour that had spread to the lung from another tissue. Even if it were a lung cancer, there are different types of lung cancer which respond differently to different treatment regimens. These additional observations could have a bearing on the type of treatment that the oncologist decides is most appropriate.

Blood Counts

Taking a blood sample is a routine procedure and provides much useful information, not only when a leukaemia is suspected. Among the things of interest are the number and type of the different cells present in the blood. Red blood cells, if low in number, may indicate **anaemia**. White blood cells are more relevant to malignant disease. In a person who is suffering with an infection their white blood cell count may be two to three times higher than the normal count. In patients with leukaemia it may be even higher and ten times the normal count is not unusual in such cases. The white cell count in leukaemia patients may also be abnormal in that immature cells, which are usually only seen in the bone marrow, may also be present.

Blood samples may be used in identifying the presence of certain chemicals which are markers for certain rare types of cancer. Elevated levels of alpha-fetoprotein (though this is also raised during normal pregnancy) are raised in patients with rare tumours of the ovary and testis, as are 5HIAA levels in patients with carcinoid tumours.

Other tests on the blood chemistry can reveal whether the urea and salt levels are abnormal, which would indicate a problem with kidney function. Similarly, changes in the levels of other chemicals may indicate abnormal liver function.

Endoscopy

The **endoscopy** procedure utilises fibre-optics technology and allows the inside of cavities to be viewed. Bundles of glass fibres, linked to lenses and contained in a flexible tube, can be guided to a point in the body which may be up to a metre away from the point of entry. It is even possible to use this apparatus to collect small biopsy samples.

When this approach is used to examine the lung, it is called **bronchoscopy**. A local anaesthetic is administered, the bronchoscope passed through the nose and into the upper air passages of the lung. It is then possible to examine all parts of the lung, usually with the patient sedated but conscious. Examination of the digestive tract can also be carried out in a similar way, except that the tube used tends to be larger and therefore the patient needs stronger sedation. **Oesophagoscopy** is used to examine the oesophagus, **gastroscopy** to examine the stomach and **colonoscopy** to examine the rectum and colon. Examination of the bladder can be achieved by passing the tube up the urethra and into the bladder. This is termed **cystoscopy**.

X-rays

There are a variety of techniques that have developed from the original X-ray procedure and have proven to be very useful diagnostic aids. Most people have had an X-ray taken of them at some time in their life. Usually this has been a chest X-ray which reveals the outline shape of internal organs such as the heart, lungs and of course bones. This type of procedure can also reveal enlarged lymph nodes in the chest that cannot otherwise be detected by routine physical examination.

The **mammogram** is an X-ray examination of the breast. It can be used to detect very small tumours, even before they are large enough to be noticed by touch. During the mammogram, the breast is pressed between two plates and exposed to a very low dose of radiation, to minimise the risk of damage. Some women find the procedure uncomfortable and there have been some concerns raised about the X-rays themselves causing an increased risk of breast cancer. However, the benefits that this procedure provides in the early detection and possible cure of the disease far outweigh these concerns.

X-ray examination of the abdomen is a useful aid in detecting problems in the digestive tract. To improve the resolution and the detail that can be seen, contrast X-rays are often taken. This usually requires a period of fasting overnight. The patient then swallows a portion of crystals followed by a bland fluid. This is called a **barium meal**. The radiologist is able to monitor the material as it makes it way down the digestive tract and can then take X-rays at appropriate times.

The most sophisticated use of X-rays is with **computed topography (CT) scanning**. In this technique the patient lies on their back and is slowly moved through a ring shaped apparatus, which has X-ray sources on one side and a series of detectors on the other. It is then possible to measure the X-ray emissions at different points around the ring, which will vary depending on the densities of the tissues that the X-rays pass through. To ensure a sharp picture the patient is required to hold their breath for a few seconds. A computer can then compare the readings at different points and relate them one to another. This enables a picture to be built up of that part of the body which lies within the ring. It is possible to move the body by as little as 0.5cm at a time and to take a scan each time. In this way any unusual swellings in internal tissues can be detected.

New Approaches to Diagnosis

The increased knowledge of mutations in genes and their association with the development of cancer has provided possible new means of diagnosing and monitoring malignant disease. A good example of this is provided by the *ERB-B2* gene, which has been found in multiple copies in the cells of many breast tumours. If there appears to be very high levels of the protein product of this gene present when a biopsy is examined by the pathologist, then there is a high probability that the disease has already spread to other parts of the body. A similar example is the overproduction of the protein coded for by the *N-MYC* gene in neuroblastomas, which is also associated with more aggressive disease. It may be that such information could lead to differences in the treatment regimen adopted by the oncologist.

The very sensitive technique of **PCR**, which allows tiny amounts of DNA to be greatly amplified so that they can be studied, has been proposed as a useful approach for identifying mutations in DNA fragments from body specimens. It has, for example, been suggested that it is possible to screen the faeces to detect the presence of mutated *P53* genes from patients at risk of developing colon cancer.

The detection of genetic mutations can also be used to monitor the efficiency of treatments given to patients. The *BCR-ABL* rearrangement, which is characteristic of chronic myeolid leukaemia, can be monitored using PCR and can detect very small numbers of surviving leukaemic cells in the bone marrow of patients after chemotherapy.

Staging

Once a diagnosis of cancer is confirmed, it is necessary to assess the stage of the tumour so that an appropriate treatment can be designed. This is a diagnostic system for determining the extent of disease and is different for different types of cancer. In essence the stage of the cancer is an indication of how widespread is the disease. Stage I indicates localised disease, whereas stage IV is associated with very widespread disease. The stages for any particular type of malignancy are fairly precise and involve internationally agreed criteria. These play a major role in helping the oncologist to decide what is the most appropriate form of treatment for a patient.

Tables 6.1 and 6.2 summarise the types of diagnostic considerations that are taken into account when staging a cancer. Table 6.1 is the

Table 6.1 Staging system used for bladder cancers.

Stage	Description of tumour
T1	Localised carcinoma
T2	Very early invasion of the carcinoma into superficial muscle of bladder
T3	Invasion of the carcinoma into deep muscle of bladder wall
T4	Invasion of organs beyond the bladder

Table 6.2 Staging system used for diagnosis of non-Hodgkin's lymphomas.

Stage	Description of lymphoma
1	Malignant cells in single lymph node region
2	Malignant cells in more than one lymph node region, but on the same side of the diaphragm
3	Malignant cells in lymph node regions on both sides of the diaphragm
4	Malignant cells in other organs

staging system used in the diagnosis of bladder cancer. The survival rates for bladder cancer vary considerably according to stage. Patients with T1 tumours have a survival rate of 70%, decreasing to 50% with T2 tumours, 20% with T3 tumours and less than 10% with T4 tumours. The Ann Arbour system is used in the staging of non-Hodgkin's lymphomas (Table 6.2). As with other staging systems the lower the stage, the greater the chance of a patient responding to treatment and being cured. In this system it is also possible to subdivide the stages depending on whether the patients show other symptoms, such as night sweats, weight loss and/or fever. Stage IA would therefore be a patient who had malignant cells in lymph nodes in only one region and had none of these other symptoms, whereas a patients who showed the other symptoms would be classed as stage IB. Patients with B symptoms have a worse prognosis than those without them. Examples of how staging can affect the type of treatment selected for some of the most common cancers are described in Chapter 8.

Chapter 7
HOW IS CANCER TREATED?

Overview of Treatment

The treatment of a patient with cancer is very much assessed on an individual basis. The first question to address is whether the patient has a chance of being cured (**curative treatment**) or at least gaining remission from the disease so that they will have several years of good quality life. On the other hand if there appears to be little chance of a cure then **palliative treatment** needs to be considered. In this case the treatments of the symptoms need to be carefully evaluated so that the patients undergo an improvement without being subjected to additional and unwanted side-effects of the treatment itself.

If it is decided that a patient may respond well to treatment, then the oncology team working with the patient will decide on the best therapy options. It is not possible to detail how this is accomplished here, as it is outside the scope of this book, but it is based on the experience of the oncology team and their knowledge of the type of malignant disease affecting the patient. The important thing to bear in mind is that the therapies are always designed with the intention of providing an effective treatment for the cancer without subjecting the patient to unacceptable side-effects and thus maximising the quality of life enjoyed by the patient.

In making these decisions it is often the case that the best chance that the oncologist has of achieving a cure, is in the first attempt at treatment. Therefore as much information as possible about the patient and their cancer is gathered and considered. Of particular concern are issues such as whether the cancer is localised or has spread to other parts of the body, whether it is likely to respond to particular forms of treatment and what its long term behaviour may be.

Surgery

Surgery was the first method used to treat cancer and in some ways is still the best option because, if the tumour is localised to one site so that it can be removed by an operation, the body will not have been exposed to any potentially damaging agents, as would happen with chemotherapy or radiotherapy. In some cases surgery alone is sufficient treatment, but in others it is more appropriate to combine surgery with chemotherapy or radiotherapy.

A patient may undergo a minor operation to remove part or all of a tumour as part of the process of diagnosing the disease. The biopsy that is removed from the patient can then be examined by a pathologist to provide information to the oncologist, first to confirm that the tumour is malignant and also to help in the assessment of the best treatment for that particular patient.

The extent of surgery that is used is very variable and depends on the size of the tumour, on the physical and mental state of the patient, and other considerations. In some cases this results in little deformity but in others there is considerable physical and functional impairment. Of course, all surgical procedures have a degree of risk associated with them and this is taken into account in deciding the best approach where more than one treatment is possible. Surgery will therefore attempt to remove the tumour mass, but may also remove nearby lymph nodes because, when tumour cells begin to spread to other parts of the body, they often first colonise nearby lymph nodes. The removal of such lymph nodes is sometimes used to help in staging of the disease.

In some cases surgery may not be regarded as sufficient to produce a cure, but it may still be regarded as a useful part of a treatment. This would occur in a patient where it is considered preferable to remove the bulk of a tumour by surgery and then treat the remaining tumour tissue by radiotherapy or chemotherapy.

The **side effects** caused by surgery are largely what most people would expect. After an operation, pain will be experienced and the amount of discomforture will depend on the type of operation and its scale. However, painkillers are available, as are anti-emetics if the patient has feelings of nausea. Patients may wake up after an operation to find various tubes (**drips** or **cannulas**) inserted into their arms. These are to aid in the administration of drugs and fluids. They are usually removed once the patient is again able to take fluids by mouth. A few patients may also have a tube (**catheter**) inserted into the bladder to drain urine, but this will not be left in place longer than is necessary. Blood and fluid may build up where part of the body has been cut and again a tube may be inserted to help this drain away. This type of **wound drain** is usually removed after a few days. If patients are unsure what is happening to them, then they should ask the nursing staff who are caring for them.

Surgery is undertaken only when the benefits outweigh potential problems. In some cases tumours may be regarded as inoperable. This means that the position or size of the tumour means that there would be unacceptable damage to other organs if surgery was attempted. This could well be the case with brain tumours. Of course surgery may mean that disfiguration or destruction of certain tissues means that **reconstruction** may be necessary. This might involve creating a new bladder, rebuilding part of the face or reconstructing a breast. The degree of reconstruction is dependent on several factors and needs to be fully discussed with the oncologist. However, the ability of surgeons to rebulit structures is very impressive, and a number of these procedures are mentioned in Chapter 8.

Radiotherapy

Almost half of all cancer patients receive some form of radiotherapy as part of their treatment. It is sometimes used alone, but may also be used in combination with either surgery or chemotherapy. Various forms of radiation are used in this therapy but they all have the same ability to destroy cells. The reason why this approach works is that the doses of radiation used are carefully calculated so that there is a selective killing of cancer cells rather than normal cells. It is perhaps ironic that one of the environmental causes of cancer, ionising radiation, is also one of the most effective ways of destroying tumour cells.

The main advantage that this approach has over surgery is that the structure and function of affected organs can be better protected than by surgical procedures. Radiotherapy appears to be particularly effective where the tumour is relatively small. As the cure rates are similar to surgery in these cases, radiotherapy may be chosen to produce a better cosmetic result for the patient.

Radiotherapy is a very technically complex approach. A number of considerations are taken into account before deciding upon the best procedure to be used including the radiosensitivity of the tumour cells and the proliferation rate of those cells. The latter is important in assessing how well the organ being irradiated will recover from the treatment, as tissues with a relatively high proportion of proliferating cells (eg., colon) will probably recover more easily than tissues with a low level of cell proliferation (eg., brain). Usually the total dose of radiation given to the patient is delivered over a period of time as a series of small doses. This approach generally appears to be a more effective way of destroying all the tumour cells. If a carcinoma is treated, then a typical course of treatment may involve giving 2 Gray of radiation every day for between three to seven weeks. The treatment of a lymphoma, whose cells are more sensitive to radiotherapy, may require only half that dose.

It is often the case that multiple beams of radiation are used. CT scanning is often used to determine the boundaries of deep tumours so that the appropriate region can be targeted. The radiation beams are focused so that they will converge in the body at the site of the tumour. This ensures that the maximum effect is produced in the tumour and the damaging effects on surrounding normal cells are minimised. The skin burns that were often observed in the past are therefore prevented.

An alternative form of radiotherapy is to insert highly radioactive elements into the tumour mass. This is not always possible but has proven to be useful for the treatment of certain tumours. For example, **caesium-137** is used in this way to treat cervical carcinoma. It is also possible to give radioactive chemicals orally if they are known to be taken up preferentially by the organ containing the tumour. For example, **iodine-131** is taken up by primarily by the thyroid and has been used to treat thyroid tumours. In these cases the highly localised source of radiation is sufficient to kill the tumour cells but causes less damage to surrounding normal cells.

Not all tumours are suitable for radiotherapy and those that are can be divided into three groups, although there is a certain amount of overlap between these groups. The first of these groups are radiosensitive tumours and include lymphomas and certain childhood tumours. The tumours of limited radiosensitivity include carcinomas of the bladder, cervix, head and neck, and skin. The final group are regarded as radioresistant but they do show some response to this treatment, which can therefore be used as part of the palliative care offered to patients with incurable disease. These tumours include colon carcinomas, melanomas and osteosarcomas.

A related procedure which is sometimes used in the treatment of leukaemias and lymphomas is whole body irradiation. This approach is designed to destroy all cells, both normal and malignant in the bone marrow, prior to the transplantation of normal donor bone marrow cells.

The use of radiation in this way can of course have side-effects, as normal cells surrounding the tumour may also be damaged. This type of damage begins to manifest itself during treatment and usually becomes fully apparent within a month after treatment has ceased. Patients treated with radiotherapy often feel unwell after treatment. This can range from feelings of lethargy to **nausea**, but this can be minimised by rest and counselling. More locally occurring effects such as **diarrhoea**, hair loss, skin rashes and **vomiting** can be treated as and when they arise. **Hair loss** is restricted to that part of the body that is being treated. In other words hair loss will occur on the chest, if the chest is being treated, but would not be lost from the head. Usually hair will regrow after the treatment is completed. The **skin burns** that used to occur are rarely seen since the introduction of modern radiotherapy machines. Sore skin might occur, but this is unlikely to be more than the equivalent of a mild sunburn. Side effects vary from patient to patient and are influenced by the region of the body being treated. Generally speaking they are temporary and will disappear.

In the longer term there is a risk of **secondary tumours** arising due to the damage of normal cells by the radiation. As these secondary tumours may not occur for 10 to 20 years after treatment this is generally regarded as an acceptable risk for a patient who would otherwise die from their present disease if not treated.

Chemotherapy

Chemotherapy has become an important weapon in the treatment of cancer and yet it produces a cure in only a minority of cases; largely in malignancies affecting children, though testicular cancer in men aged between 20–40 is also often cured by chemotherapy. In cases where the tumour has metastasised and formed numerous secondary tumours in different regions of the body, surgery and radiotherapy may be impractical and therefore chemotherapy is the only option. The prime aim for chemotherapy in treating the common solid tumours which occur mainly in middle to old age is to produce a remission from the disease. This can provide for relief of symptoms and occasionally disappearance of the cancer. However, the tumour generally returns and the patient ultimately dies from the disease.

Table 7.1 summarises the responses of a number of tumours to chemotherapy. A small number of tumours, such as certain lymphomas, testicular cancer and acute leukaemias, are susceptible to chemotherapy so that after treatment the patients frequently enter long term remission and are effectively cured. Some tumours are highly responsive to chemotherapy and the disease can be controlled for many years by chemotherapy. Patients with low grade non-Hodgkin's lymphoma may survive for over 10 years before their disease becomes resistant to treatment. Tumours, such as bladder cancer, that are moderately responsive to treatment may be controlled for a year or two before failing to respond further. Tumours such as those of the brain and kidney are only slightly responsive and show little or no response to chemotherapy.

Table 7.1 Susceptibility of different cancer types to chemotherapy.

Susceptible	Highly responsive	Moderately responsive	Slightly responsive
Acute leukaemias	Breast cancer	Bladder cancer	Brain cancer
Childhood malignancies (majority)	Chronic myeloid leukaemia (chronic phase)	Cervical cancer	Renal cancer
High grade non-Hodgkin's lymphomas (minority)	Low-grade non-Hodgkin's lymphoma	Colon cancer	Liver cancer
Hodgkin's disease	Osteosarcoma	Head and neck cancer	Melanoma
Testicular cancer	Ovarian cancer	High grade non-Hodgkin's lymphomas (majority)	Oesophageal cancer
Trophoblastic disease	Multiple myeloma	Non-small cell lung cancer	Pancreatic cancer
	Small cell lung cancer	Prostate cancer	
		Soft-tissue sarcoma	
		Stomach cancer	

There is of course a certain amount of variability between individuals with each of these tumours.

Chemotherapy is likely to be most effective if the tumours in a patient are few and small. As a patient with advanced disease may have 10,000,000,000 cancer cells, then even removing 99.9% of them by chemotherapy still leaves large numbers that can survive and acquire further mutations which make them resistant to subsequent treatment. Therefore chemotherapy is often combined with surgery or radiotherapy, which are first used to reduce the tumour cell burden before drugs are administered.

The cytotoxic drugs that are used in chemotherapies are often given to the patient in a series of administrations over a three to four week period. This approach appears to be preferable because it reduces the risk of drug resistance occurring and appears to cause less damage to normal cells. As with radiotherapy, the drugs that are used in chemotherapy damage both malignant and normal cells. However, they produce a positive effect because they cause more damage to the malignant cells. They do this by selectively killing cells that are in the cell cycle (see Chapter 2) and, as more of the cells in a tumour tend to be proliferating than in normal tissues, then the tumour cells tend to be more susceptible to chemotherapy. There are some exceptions to this. For example, tumours may have fewer proliferating cells than are found normally in the stem cell populations of bone marrow, gastrointestinal tract or skin. However, these stem cells appear to be less susceptible to cytotoxic drugs than the tumour cells and are certainly far more efficient at repairing DNA damage. Nevertheless, all of these normal cell populations are damaged to some extent by chemotherapy. The reduced numbers of white blood cells and platelets in the blood can be regarded as an indicator of the amount of damage the bone marrow has suffered. The white blood cell and platelet counts are therefore used to judge when the next administration of cytotoxic drugs is appropriate. The recent introduction of growth factors which stimulate the bone marrow cells to proliferate has proven a useful tool to increase blood cell numbers and to shorten the period of time between cytotoxic drug administration. The shorter the time, the less chance the tumour cells have to recover their numbers and to become drug resistant.

Cytotoxic drugs do kill cancer cells and by adjustment of their administration can kill them relatively selectively compared to normal cells. The major problem with this form of treatment is that cancer cells do eventually become resistant to these drugs. The problem of drug resistance is at the heart of achieving better cure rates and it is likely that this is intimately associated with blocks in apoptosis (see Chapter 3).

There are a number of side effects with this type of treatment. As described above, chemotherapy can kill a number of bone marrow cells, which in turn will reduce the numbers of blood cells. The first problem with this, is that the immune response of the patient will be compromised and therefore care has to be taken to avoid infection, which might otherwise overwhelm the patient. Haemorrhaging is a problem which arises from a very low platelet count in the blood. This does not appear to be a significant problem in patients being treated for solid tumours but can occur in patients being treated for leukaemia. Long term problems related to anaemia may also occur in some patients. To overcome these problems there has been a recent introduction of bone marrow cells growth factors which stimulate the stem cells to proliferate and so replenish the blood cell numbers more quickly. It is also possible to remove bone marrow cells from patients and store them until after the chemotherapy. These cells can then be reintroduced into the patient and again help the restoration of blood cell counts.

Nausea and vomiting are common side effects, but again the recent introduction of anti-emetic drugs has allowed vomiting to be better controlled. There are other, perhaps less effective, drugs which can help in alleviating longer term problems with vomiting.

Although cancer is primarily a disease of middle to old age, it is possible that some patients may wish to have children in the future. This can therefore be an important issue as the germinal cells are particularly sensitive to chemotherapy and contraception is necessary during treatment to avoid spontaneous abortion or abnormal embryo formation. In fact it is recommended that the patient is not involved in conception for at least a year after treatment has ended. As an added precaution against chemotherapy-induced sterility, male patients are usually offered the opportunity of having their sperm cryopreserved.

Hair loss is often regarded as a matter of major concern, particularly by women. Not all chemotherapies cause hair loss and some only produce a thinning of the hair which is not really noticeable. In some specialist oncology hospitals, machines are available which cool the scalp during treatment, which restrict blood flow to the scalp and protect the hair follicles against exposure to the chemotherapeutic drugs. This is not always applicable, but in some cases can prevent hair loss. When hair loss does occur, it is always temporary and the hair will grow back after treatment is completed. The oncologist can often predict if a particular drug will cause hair loss and advise the patient in advance of treatment. Wigs, hats and headscarfs can all be used by patients until their hair regrows.

It is difficult to predict exactly how different patients will react to different chemotherapies. Some patients appear to cope well with this type of treatment, whereas others experience major difficulties. Certain drugs have specific side effects and can result in damage to particular organs. Long term exposure to methotrexate can produce damage to the liver. Regimens involving bleomycin can cause lung damage and respiratory problems. Administration of anthracycline drugs has been associated with heart failure, cisplatin often causes kidney damage and vincristine can cause damage to the nervous system. These side effects reinforce the fact that chemotherapeutic agents are cytotoxic and need to be used selectively. The prevention of cancer is a far better strategy than trying to cure it.

Other Treatments

It is now accepted that the growth of certain tumours is hormone dependent. This has an interesting implication because it suggests that altering hormone levels in the body might cause a tumour to regress. This **hormone therapy** rarely produces a cure but can control malignant disease for many years. It is often used in conjunction with other types of therapy, that is surgery, radiotherapy or chemotherapy. Hormone therapy has been most successfully used in the treatment of breast and prostate cancers. In breast cancer the use of the anti-oestrogen, tamoxifen, is frequent.

   Immunotherapy is based on the idea that it may be possible to trick the body's own immune system into attacking the body's cancer cells. The fact that this might be possible is demonstrated by the observation that AIDS patients (who have a compromised immune system) often develop unusual forms of cancer. If the immune system normally destroys such cancers then maybe it could be persuaded to attack more common tumours. There has been an enormous amount of research in this field for many decades, with many high hopes, but so far this has not produced any significant improvements to anti-cancer therapy.

Palliative Care

In those patients where the disease appears to be incurable, the question of palliation arises. **Palliative care** has been defined by the World Health Organisation as the active total care of patients whose disease is not responsive to treatment. Control of pain, of other symptoms and of the psychological, social and spiritual problems, is paramount. The goal of palliative care is the achievement of the best quality of life for patients and their families, so that it produces the best balance between quality of life and length of life for each patient. One of the most important aspects of palliation is counselling, as many of the symptoms in terminal illness can be frightening for the patient and worrying for their relatives. It is important to note that a third of patients need not remain in a **hospice** but can be managed by a hospice providing day care.

Pain management requires an accurate assessment of the cause of the pain and determination of whether it is possible to eradicate the source of the pain. It is often the case that pain killing drugs will have to be administered. There are three general categories of pain, ranging from mild to moderate to severe. For mild pains, non-opioids such as paracetamol are used. Moderate pains require a weak opioid such as codeine and for severe pain increasing doses of strong opioids, such as morphine are prescribed.

There are diverse symptoms associated with terminal disease, which all have possible treatments to help alleviate their effects. Surgery may be employed as part of palliative care. For example, a patient with a large colon tumour may have a colostomy to relieve an obstruction, or a patient with laryngeal cancer may have a tracheotomy for similar reasons. Similarly palliative radiotherapy may also be given, but this would use doses of radiation at much lower levels and for fewer times than those used in radical radiotherapy.

Why do Cancer Patients Die?

Patients with terminal cancer may die from a variety of causes, some direct and others indirect. In fact many patients die of an unrelated condition such as heart disease, infection or even an accident. Patients often develop a decreased resistance to infection so that terminal bronchopneumonia or infection of the urinary tract may be fatal. In many cases it is not possible to establish the immediate cause of death. Where the cause of death is directly related to the disease, it is usually due to the spread of the cancer to a vital organ such as the brain, liver or lung.

What are the Chances of a Cure?

The chances of a cure depend on a number of factors. Early detection is certainly of great importance and the need for self-awareness as far as changes in one's own body is concerned cannot be understated. The prognoses for different types of cancer can vary considerably and therefore the chances of a cure are dependent on the type of cancer that occurs.

What this means is that a cancer that is early in its development, localised to one site in the body and can be treated by surgery or radiotherapy is most likely to be cured. If the disease has begun to spread or fails to respond to radiotherapy, and as mentioned above some tumour types do not respond well to this type of treatment, then chemotherapy needs to be used. The chance of a cure after chemotherapy is dependent on the tumour type being treated. Table 7.1 summarises some of the main types of cancer and their response to chemotherapy. Patients with tumours in the susceptible category are often cured in this way. Patients with other tumours may not be cured but can gain many years of good quality life which they would not have had without treatment.

The important fact is that treatments for cancer are continually evolving. As the years go by, the treatments become more effective. It is important for a patient to be positive and for oncologists to seek remissions from disease; in the extra years that the treatments provide you, further improved techniques may become available. This all helps to stack the odds in your favour.

Chapter 8
TREATMENTS FOR COMMON CANCERS

Introduction

The previous chapter dealt in general terms with the types of therapy that are used for the treatment of cancer. This chapter will focus in a more specific way on the most common types of cancer and the usual approaches adopted by oncologists in dealing with them. There is always variation in the type of treatment provided, as much depends on factors which vary from patient to patient. Also treatments at specialist cancer hospitals also frequently involve patients being entered into clinical trials for new drugs or procedures. Nevertheless the following summaries provide a general guide to the type of treatment a patient with one of the more common types of tumours (Table 1.1) can expect to receive.

One of the interesting features of cancer treatments is that the response of the patient to therapy can vary considerably depending on the type of cancer being treated. In assessing the effectiveness of treatment, the oncologist usually regards **survival** for more than five years after first treatment to be an indicator of good prognosis. Examples of the proportion of patients with different cancers surviving for more than five years are shown in Figure 8.1. This study from the USA demonstrates that not all cancers have the same susceptibility to treatment; prognosis for a patient with lung or pancreatic cancer is likely to be poor, whereas patients with breast or prostate tumours may survive for many years after diagnosis. This study also indicates that survival rates in the USA are often higher than those observed in other countries. For example, five year survival for prostate cancer is observed in over 80% of patients in the USA, but in the United Kingdom this figure is nearer 40%. Similarly, five year survival for breast cancer is attained by over 80% of patients in the USA but in less than 70% of patients in the United Kingdom.

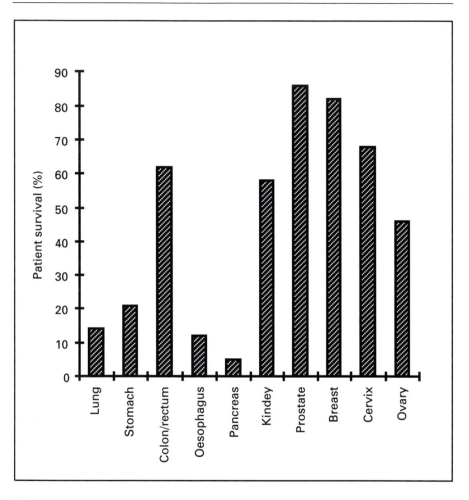

Figure 8.1 Cancer survival rates in the USA. The estimated percentage of patients with different types of cancer who survive for more than five years after diagnosis.

Lung Cancer

Treatment depends to a certain extent on which of the four types of lung cancer the patient has. Squamous cell carcinoma makes up half of all cases and has its origins in the cells that line the airways. Adenocarcinoma of the lung accounts for 20% of cases and originates from cells that produce mucus in the lungs. Large cell carcinoma comprises 10% of cases and with the other two types mentioned above are collectively known as non-small cell lung cancers. The remaining 20% of lung cancers are of the small cell type.

If non-small cell cancers are diagnosed at an early stage they are less likely to have spread and therefore could be treated by surgery. As long as the tumour is localised to the lung, then surgery to remove part or all of the lung gives a reasonable chance of a cure. About a quarter of people with these tumours are regarded as being suitable for surgery. The others are not offered this possibility because their tumour has already been found to have spread, or they are insufficiently healthy for an operation. Occasionally a patient who has a small tumour, which does not appear to have spread, but who is not fit, may be offered radiotherapy. The use of radiotherapy in conjunction with surgery may produce improvements to the survival rate. For advanced disease, chemotherapy is the main option. It offers the possibility of reducing the tumour size, of controlling symptoms for some time and providing a prolongation of life. However, it is unlikely to be curative.

The small cell cancers are usually more aggressive forms of lung tumour and surgery is rarely used in their treatment because they have often spread by the time that they are diagnosed. Chemotherapy is therefore the treatment of choice. This often produces a prolongation of survival and in a few patients even a cure. Radiotherapy is also given to those patients where the cancer does not appear to have spread beyond the chest.

A rare form of lung cancer which has been in the news because of its association with exposure to asbestos is **mesothelioma**. This tumour may take up to 40 years to develop after the patient's initial exposure to asbestos. It originates in the cells forming the lining of the lungs and if detected early can be cured by surgically removing the linings; however, in most cases a cure is not possible.

Approximately a quarter of patients with lung cancer who have had surgery are cured, though only a very few patients with small cell carcinoma show evidence of a cure. Unfortunately the majority of patients with lung cancer still die within two years of diagnosis.

Stomach Cancer

The most common way of treating stomach cancer is by surgery. If possible only a part of the stomach will be removed, though if this is not possible then the entire stomach will be removed and the oesophagus reconnected with the small intestine. The local lymph nodes are also removed in these operations to check for the spread of tumour cells. This operation will have implications for the eating patterns of patients who have had their entire stomach removed (a complete **gastrectomy**). They may also need to have vitamin B12 injections, as this vitamin needs the stomach to aid its absorption into the body.

If the lymph nodes are found to contain cancer cells, then it is likely that the tumour cells may also have spread to other parts of the body. In this case chemotherapy can be used to reduce tumour masses and to control the disease for a time, but the patients inevitably die.

If the cancer is identified at an early stage then surgery can produce a cure in over half of the cases. However, in over 80% of patients the disease is not identified at a sufficiently early stage and will already have begun to spread to other tissues.

Breast Cancer

Most women with breast cancer have surgery to remove the tumour. Surgery is generally not as drastic as it used to be. The survival rates of women appear to be similar whether they have smaller surgical operations (**lumpectomy**) or removal of the breast (**mastectomy**). The ability to carry out a breast reconstruction has improved in recent years and this has important psychological benefits to the patient. It is usual for surgery to be followed by radiotherapy to destroy any remaining micrometastases which may have escaped the surgery. Both the area of the breast and the armpit, where the local lymph nodes are found, are treated.

It is often the case that hormone treatment or chemotherapy may be given after surgery in an attempt to destroy any tumour cells that may have spread to other tissues. If metastatic spread is obvious then these approaches will certainly be followed. Hormone therapy usually involves the use of tamoxifen, which works by preventing oestrogen being taken up by oestrogen receptors inside the cancer cells. This effectively acts as a block on the mechanism that encourages the cells to proliferate. It is usually women who are postmenopausal who are given this treatment, but some premenopausal women with advanced disease are also given **tamoxifen**. One treatment for premenopausal women is to give drugs which prevent the release of hormones by the pituitary gland. This has the effect of inhibiting the release of oestrogen by the ovaries.

Chemotherapy is usually used in premenopausal women with advanced disease, particularly if hormone therapy no longer appears to be effective.

Patients with breast cancer can be cured but again the chances of a cure depend on how soon the tumour is detected. If they have localised disease then there is a high chance of a cure. Even if the tumours reappear they tend to do so locally and can be controlled for many years. The cure rate drops considerably in patients with metastatic disease, but again in some women the secondary tumours can be controlled for many years.

Colon and Rectum Cancers

Surgery is the main type of treatment and the aim would be to remove the tumour and a surrounding area of apparently normal tissue which usually allows for the two remaining ends of the colon and/or rectum to be reconnected. As usual in these operations, the local lymph nodes would also be removed to check for evidence of spread of the tumour cells. If the two ends of the bowel cannot be rejoined, possibly because the region to be removed was near to the anal sphincter, then a **colostomy** would be required. This involves bringing the open end of the bowel out onto the surface of the abdominal wall. A bag is then worn over the opening to collect the faeces. In some cases this is a temporary situation which can be rectified by another operation, but it may be a permanent arrangement. With advances in surgical technique, this is becoming a less common procedure.

Chemotherapy may also be given if there are reasons to believe a relapse may occur, as would be the case where tumours cells were detected in local lymph nodes. In the cases where the tumour arose in the rectum, radiotherapy may also be given to the pelvic region.

Surgery has a very high chance of producing a cure if the cancer is detected an early stage, with perhaps more than a half of these patients being cured. However, about half of all patients with bowel tumours are not diagnosed until their disease has reached an advanced stage. Like other tumours, these show the best cure rates if detected early and the more they have spread, the poorer the prognosis.

Cervical Cancer

One of the most important breakthroughs in cancer prevention in recent decades has been the introduction of the **Pap smear** test (see Chapter 10), which can allow for the identification of premalignant changes in cervical epithelial cells. The regular screening of women means that action can be taken to treat these conditions before they transform into fully malignant disease. These precancerous cells can be destroyed by either freezing, heating or laser treatment. Alternatively they can be removed by minor surgery. The cure rate in cases showing these early signs of disease approaches 100%.

If carcinoma of the cervix develops and is localised then this is usually treated by surgery, which involves a hysterectomy and removal of local lymph nodes. The ovaries are generally not removed, but if this proves necessary in premenopausal women, they are subsequently given **hormone replacement therapy**.

Although radiotherapy appears to be as effective as surgery in early stages of cervical cancer, it has worse side effects including loss of ovarian function, and therefore surgery tends to be used. However, if the cancer has spread from the cervix then radiotherapy is often used. In some cases where the cancer cells are found in local lymph nodes, radiotherapy is used after surgery. Radiotherapy can be given both externally and internally. The internal treatment is provided by inserting an applicator into the cervix which contains a radioactive source. The applicator is kept in place for one to two days, while the patient remains in bed.

Chemotherapy is given to women who either have disease that has spread to other parts of the body, or who have relapsed after radiotherapy. The most effective chemotherapeutic agents for this tumour appear to be based on platinum.

The chances of a cure are very high if the disease is detected in its early stages. Even when it has spread locally there is a chance that radiotherapy will result in a cure in some women and certainly prolong life in others. The more advanced the disease the less chance of a cure and it is in these circumstances that chemotherapy is useful in controlling the disease for a time.

Prostate Cancer

The prostate is a small gland at the base of the bladder which produces fluid in which the sperm swim. The first signs of prostate cancer can be detected by a blood test for the prostate specific antigen, but in only 20% of positive results do men go on to develop life threatening cancer.

In men with localised disease there are a choice of treatments depending on the precise nature of the patient's tumour. Radical surgery may be used, though this is technically difficult and may result in impotency. Radical radiotherapy may be used where surgery is not possible, though it appears to give less satisfactory results and can lead to complications involving the bowel and bladder. In some cases, particularly where the tumour appears to be slow growing, it may be felt best to simply wait until more acute symptoms are observed.

Where the cancer has spread more widely, hormone treatment is used. Male hormones have been found to support the growth of these cancer cells. Therefore ways of reducing the circulating male hormone levels are used. Castration has been used extensively in the past although hormone treatments are now more commonly used. These involve using drugs which either block the effect of the male hormones or block their release. Chemotherapy appears to be of limited value in treating this tumour.

The control of symptoms can be achieved for many years, especially where the tumour is localised, though this tumour is responsible for 10% of all cancer deaths.

Liver Cancer

Surgery is not often used to treat liver cancer. In theory a single tumour could be removed by surgery and then the liver's own amazing powers of regeneration could restore it to its original size. However, it appears that by the time liver cancer is usually detected it is characterised by multiple tumours, which have spread from a primary tumour.

Radiotherapy is not often used as normal liver cells appear to be very sensitive to radiation. Chemotherapy is also of limited value.

The prognosis for liver cancer is not good and most people who develop this tumour die from it, with only a few being cured by surgery.

Oesophageal Cancer

If the cancer is localised to the oesophagus then surgery is the preferred treatment. After removing the affected region, the remaining ends can be reconnected. If a large piece of the oesophagus has to be removed then a piece of intestine can be used to join the remaining oesophagus to the stomach.

Radiotherapy has been used to treat cancers that have spread from the oesophagus to other tissues. This has some value in controlling the disease. However, better outcomes have been produced when radiotherapy is combined with chemotherapy. The latter often involves drugs such as fluorouracil and cisplatin.

Early detection and surgical removal of a localised tumour usually produces a cure. However, more advanced tumours can respond well to combined surgery, radiotherapy and chemotherapy, which sometimes produce a cure.

Skin Cancer

Surgery is the main form of treatment for skin cancers. In fact in many cases the surgery used to remove a suspect tumour for biopsy is all the treatment that is needed. These tumours may also be removed by freezing with liquid nitrogen or by electrocauterisation (the use of an electric current to cut out the tumour and at the same time cauterise the area).

Melanomas are potentially more dangerous forms of cancer. In this case the entire tumour is removed, as is part of the surrounding normal tissue. If the area removed is large then plastic surgery may also be required. It may be necessary to remove local lymph nodes to determine whether cancer cells are present. The removal of lymph nodes would of course require the use of general anaesthetic.

Radiotherapy is also very effective against basal cell and squamous carcinomas and is often used instead of surgery where treating tumours on the face could cause unwanted scarring. It is not used for melanoma.

Chemotherapy may be used to treat the disease when it has spread to other parts of the body and is most useful for relieving symptoms and reducing tumour mass.

Skin cancers are curable if caught early enough. This is also true of melanomas. If people are aware of their own bodies then it should be possible to detect any unusual changes, though these tumours can occur in some unexpected places, such as the soles of the feet and top of the head. However, if melanoma has reached an advanced stage and started to spread it is very dangerous.

Bladder Cancer

Diagnosis of a tumour is often accomplished by cystoscopy (Chapter 6) which not only enables the cancer to be seen but also allows for a biopsy sample to be obtained. Treatment depends on how far the cancer has spread and staging of the tumour (Table 6.1) is important in this regard. Superficial tumours that are restricted to the lining of the bladder are usually treated by surgery, though destroying them by use of a laser has also been used experimentally. However, as these tumours are potentially unstable and may have spread to other parts of the bladder lining, the patient needs to be re-examined at future times. The majority of patients with these types of tumours survive for more than five years.

Where tumours have begun to invade the wall of the bladder, it is usually necessary to remove the affected area, or possibly the whole bladder, by surgery. In the latter case, a replacement (**urostomy**) needs to be constructed (usually from a piece of intestine) which drains out through the wall of the abdomen. This may require a period of adjustment before the patient becomes used to it. Depending on the size and the spread of the cancer, radiotherapy may be used instead of, or in addition to surgery. In some cases chemotherapy can be used to reduce the size of the tumour before it is removed by surgery. The chances of surviving invasive tumours are very much lower than in patients with superficial tumours.

Ovarian Cancer

Surgery is the primary treatment for ovarian cancer. It involves carrying out a hysterectomy, removal of both ovaries and fallopian tubes. Any other malignant tissue that is noticed by the surgeon in the abdominal cavity is also removed.

Chemotherapy may also be administered to destroy any secondary tumours that are present in the abdomen (ovarian cancer does not usually spread outside of the abdomen). Usually drug treatment is based on platinum and can be quite effective. One of the more recently introduced drugs, **taxol**, has also been found to have encouraging effects on ovarian cancer. Radiotherapy is of limited use in treating this cancer.

Detection of the tumour at an early stage and treatment with surgery and chemotherapy often produces a cure. However, in more advanced stages cure rates are much lower, but chemotherapy is of value in prolonging good quality life.

Lymphoma

The lymphomas are a very disparate group of malignancies but have been classed into the general categories of Hodgkin's disease and non-Hodgkin's lymphomas. The latter are further sub-divided into low and high grade disease.

If Hodgkin's disease is localised then radiotherapy is often used and produces very good results. Usually the radiotherapy is given in bursts over a number of days for several weeks. Combination chemotherapies have also proven to be effective in treating Hodgkin's disease and are useful for more widespread disease. More than half the patients with Hodgkin's disease are cured and many others gain prolonged good quality life.

Low grade non-Hodgkin's lymphoma is an indolent disease which can be controlled for several years by oral chemotherapy, such as chlorambucil. However, these tumours often transform into drug resistant high grade disease, which is usually fatal despite more intensive chemotherapy being administered. Some patients are initially diagnosed with high grade disease. These patients are treated with combination chemotherapy, such as in the CHOP regimen (cyclophosphamide, vincristine, prednisolone). Many patients show little response to treatment or quickly relapse and die. However, up to half show a remarkable response to treatment, enter long term remission and are effectively cured.

Leukaemia

Leukaemia by its very nature is a disseminated disease and therefore surgery is inappropriate. The treatment of choice is chemotherapy. The chemotherapies that have been developed are designed to attack the leukaemic cells in different ways so that hopefully they will destroy them without destroying all the normal cells in the bone marrow. This will then enable the bone marrow to regenerate a new population of normal white blood cells to repopulate the blood. A variety of treatment regimens have been developed for the many types of leukaemia.

For acute leumaemias, drugs are usually administered at intervals of a week or more. During this time the blood will be monitored both for the presence of malignant cells, but also for other normal functions. For example, if the number of red blood cells become too low then it will produce anaemia. When the blood sample looks normal a bone marrow sample is often taken and if that is also normal then the patient is classed as being in remission. In some children this treatment results in a cure but in other children and in many adults the disease will eventually recur.

To deal with the risk of recurrence, bone marrow transplantation is sometimes considered. In this case the patient's own bone marrow is first destroyed by chemotherapy or by radiotherapy. It is hoped that this will destroy all remaining malignant cells. New bone marrow cells from a genetically related donor can then be transfused into the patient and, if they are not rejected by the patient's body, will eventually repopulate the body with normal blood cells.

Chronic myeloid leukaemia is conventionally treated with oral administration of drugs such as busulphan, which can be used to control the disease for several years. However, after a few years patients usually undergo a blast crisis, in which the leukaemic cells become more aggressive and their number rapidly proliferate in the blood. This phase of the disease is drug resistant and invariably fatal. Chronic lymphocytic leukaemia is a mild form of malignancy occurring mainly in the elderly. In many cases no treatment is necessary and the affected individuals may die of other causes rather than this disease. If some of the symptoms begin to cause problems they can usually be controlled by oral administration of chlorambucil or prednisolone.

Section II

How To Stack The Odds In Your Favour

Chapter 9
CANCER AND LIFESTYLE

Introduction

There are now many treatments available to aid the fight against cancer, and there is often much debate as to which therapeutic regimen might offer the best hope for a particular patient with a particular type of cancer. However, there is one thing that all oncologists are agreed upon and that is that the best way to avoid dying from cancer is to avoid developing cancer in the first place. The rest of this book is devoted to explaining how it is possible to adopt a lifestyle which greatly reduces the risk of developing cancer. There are some basic facts which need to be appreciated to adopt successful strategies.

The first fact is that we cannot predict any lifestyle that will offer a guarantee of complete freedom from cancer. However, there are a number of informed decisions that an individual can make which will greatly affect the chances of them developing cancer. This is about probabilities and many people clearly have difficulties in understanding probabilities. Consider for example a person who buys a lottery ticket where the probability of winning may be only 1 in 14,000,000, or a person who decides not to eat beef because there may be a probability of 1 in 300,000 that they will develop a variant form of Creuztfeldt-Jacob disease. Such people have obviously considered the chances and have concluded it is in their interest to buy a lottery ticket or to stop eating beef. On the other hand those same people may smoke cigarettes even though the probability of developing cancer as a result of this behaviour is 1 in 4. There seems to be a lack of rationality in such decisions, and it is important that the reader carefully considers the relative importance of each of the risk factors. Although a group of factors may be identified as increasing the chances of developing cancer, it is important to consider how much of a threat they each constitute in turn. There is little point in deciding to do something about a factor which carries a small risk but ignore a factor which carries a major risk. It is also important to remember that when a body is exposed to two cancer causing risks then the cumulative effect can be greater than that posed by the risks individually.

A second fact to consider is that of the complexity of cancer and the multiple influences that combine to create a fully malignant cell. Most cancers are the result of several different causes and only rarely will one factor be responsible for all of the damage required to transform a cell into a cancer cell. An example of the latter might be where individuals are exposed to very high radiation doses as a result of close proximity to an atomic bomb explosion or large radiation leak. Generally developing cancer is about being subjected to a number of influences over a period of time. Therefore, one should always consider the spectrum of influences that are present. There are genetic factors and a whole range of environmental factors, both protective and causative, which can combine to increase or reduce the risk of developing cancer. In some cases we can state that there is a definite link between a factor and some form of cancer, for example smoking and lung cancer. In other cases we claim a probable link, for example high salt intake and stomach cancer and in some cases there may be a possible link, for example, not breast feeding and breast cancer. So it is important to consider relative risks and not perceive all risks as being of equal significance.

I think a good way of looking at this situation is by using the analogy of the human body as a castle which is under siege from attacking forces. The strength of the castle depends first on the strength of its walls' foundations, which can be regarded as the genes in the body. Good quality genes, or strong wall foundations, act as the best defence against attack. However, if defective genes, or faulty foundations, are present, then this is a cause for concern. It means that in such circumstances it is important to be particularly aware of the predisposition to damage which can result from these defects. Most of the problems facing the body, or the castle under siege, come from outside. The attacking army that can undermine the health of the body include chemical agents (present in the diet and air), physical agents (sunlight) and viruses. By looking from the walls of your castle, which is your body, the attacking forces can usually be seen and you have it in your own hands to prevent them from gaining entry. There will always be some attacking forces out there, throughout your life, and so a constant vigil is required.

Evidence that Altered Lifestyle Changes Cancer Risk

Some of the strongest evidence that lifestyle is important in determining the risk of cancer has been provided by studies of migrants. The types and incidence levels of different cancers varies in different regions of the world. It has therefore been informative to study how these observations change when a group of people from one population move from one region and join a different population in another region.

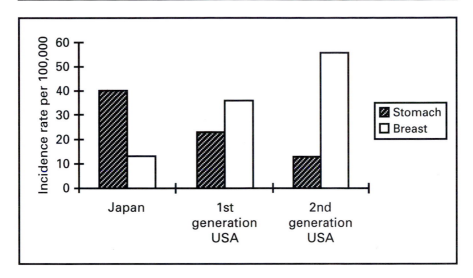

Figure 9.1 Incidence of breast and stomach cancers in Japanese women in Japan and the USA.

During the twentieth century there has been a significant movement of Japanese people to the USA and it has therefore been possible to observe how the cancer rates in first and second generation women of Japanese origin change with time in the USA. The incidence of breast cancer is of particular interest because of the relatively low levels of this tumour in Japan and the high levels found in America. Figure 9.1 demonstrates that breast cancer is more common in Japanese women after they have moved to America compared to Japanese women who live in Japan. Interestingly, second generation women who were born and grew up in America have even higher rates of breast cancer, which approach those observed in white American populations. What is even more interesting is that the incidence of stomach cancer in these groups shows the reverse trend, with it becoming less common, the longer the Japanese women have lived in the USA.

A similar study has been carried out on Chinese men by comparing the incidence of stomach cancer and prostate cancer in groups in different parts of the world. Figure 9.2 shows that the incidence of stomach cancer is very high in China, less so in the more developed Singapore and lowest in the USA. Whereas the reverse is found for prostate cancer.

These types of observations indicate that the development of cancer is under external environmental influences and is not necessarily determined by racial characteristics. They also indicate that risks of

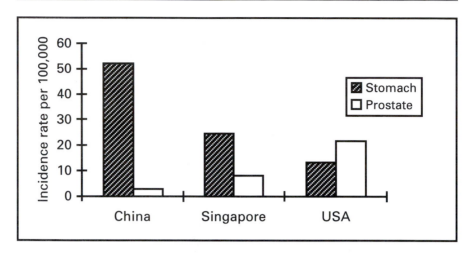

Figure 9.2 Incidence of prostate and stomach cancers in Chinese men in China, Singapore and the USA.

developing cancer can be changed dramatically in a relative short period of time depending on the external influences that a body is exposed to.

Occupational Risk

It is probable that today in developed countries there is only a small proportion of cancers that can be attributed to occupational risk. This subject is of interest though, as one of the earliest studies linking cancer to the environment involved an investigation of the occurrence of a particular type of cancer in workers exposed to a factor associated with their work. In the late eighteenth century the British doctor Percival Potts noted that there appeared to be a high incidence of cancer of the scrotum in a number of young men. Upon later investigation he was able to establish a link between this tumour and the work of his patients as chimney sweeps. The daily exposure of small boys to the soot in the chimneys of London houses appeared, over several years, to lead to the development of this tumour. Potts concluded that there was one or more chemicals in the soot responsible for the development of the tumours.

It was subsequently observed that bladder cancer was common in workers in the dye and rubber industries. This was later shown to be due to exposure to a number of the chemicals used in these companies. These fell into a group of chemicals known as aromatic amines and in one case the use of the chemical 2-naphthylamine resulted in all 19 workers in one factory developing bladder cancer.

Perhaps the greatest risk to health has occurred in factories which used asbestos. This has been found to be a potent cause of cancer, particularly an otherwise rare form of lung cancer called mesothelioma.

Nowadays the potential of cancer inducing chemicals in industry is well known in developed countries and government health agencies are able to ensure that appropriate working practices are followed. Whether the same can be said for some of the developing Third World countries, where the imperative is often for industrial growth, is not so clear.

Sex

Sex is usually involved with most things in life, and cancer is no exception. This has been observed both directly in the case with cervical cancer, as well as more indirectly with lymphoma and Karposi's sarcoma in AIDS patients.

The involvement of human papilloma virus in causing cervical cancer is discussed in more detail in Chapters 5 and 14. It is noteworthy that the cancer causing properties of this virus are primarily observed in women, where it influences the development of cervical cancer. It appears therefore that cervical cancer can be regarded as a sexually transmitted disease, in which men are primarily carriers of the disease.

AIDS is another sexually transmitted disease, this time involving human immunodeficiency virus (HIV). This virus does not in itself cause cancer, but one of its effects is to allow certain unusual types of cancer to emerge. HIV destroys white blood cells which comprise part of the body's immune defence system. These white blood cells normally function by destroying invading bacteria and viruses, a function at which they are very proficient. It appears that they will also remove mutated body cells that have become changed to such an extent that they are no longer recognised as belonging to the body. These include some unusual types of cancer cells. In a person who is not infected with HIV, these white blood cells therefore destroy a number of cancer cells, but when they are themselves destroyed by HIV, then these very abnormal cancer cells can survive in the body to form tumours. The most common types are an unusual, very aggressive, lymphoma and what is normally regarded as the rare Karposi's sarcoma.

There has been some debate in recent years about the risk of cancer in women using the contraceptive pill. Although there initially appeared to be some evidence for such a risk, this has not been supported by more recent studies. At present we therefore have to conclude that women taking the pill are not at any greater risk of developing cancer.

Breast Feeding

The question of breast feeding is an interesting one, because previous studies into risk factors associated with breast cancer have indicated that women who have children and breast feed them are less likely to develop breast cancer than those who do not breast feed. The more frequently a woman breast feeds babies, the more protective the effect.

There are of course other important issues concerned with breast feeding. There is no doubt that babies gain more nourishment from mother's milk than from a formulation, and most importantly can gain immunity to pathogens during the first few months of life until their own immune system is fully functioning. These are particularly important issues in the developing countries of the world. There is also the fact that breast feeding can reinforce psychological ties between the mother and the baby.

Of course in the industrialised countries where many women work, it can be difficult to organise the breast feeding of babies, in which case women should pay particular attention to other risk factors that influence the onset of breast cancer.

An interesting observation that is often mentioned in regard to this and the above section concerns the cancer incidence in nuns. It appears that these women have a much lower incidence of cervical cancer, but a relatively higher incidence of breast cancer, than women in the general population. It has been suggested that this is evidence for the association of cervical cancer with HPV infection caused by sexual intercourse, which presumably is less frequent in nuns than in other populations of women. However, the lack of child bearing and breast feeding activities in nuns results in the higher incidence of breast cancer.

Hormone Replacement Therapy

Hormone replacement therapy (HRT) is used to treat unwanted symptoms that occur in women at the time of the menopause. It also may have beneficial effects on preventing heart disease and osteoporosis (weakening of the bones). The benefits of this treatment in many women are beyond doubt, but some concerns have been raised concerning the possibility that it may increase the risk of cancer. The situation is still under investigation but there may be a slightly increased risk of breast cancer; though this is still a matter of debate. The risk of developing either cervical or ovarian cancers are no greater than in women who do not take HRT. However, there is evidence that women who are taking oestrogen-only HRT may be at a slightly increased risk of cancer of the uterus.

Chapter 10
YOUR BODY AND CANCER
AWARENESS

Introduction

There is a small group of cancers for which there is a clear association between inherited genetic traits and the risk of developing cancer. Tumours such as these are in fact very rare but their strong genetic associations provide powerful support for the idea that genes can play an important role in the development of cancer. In some cases a single variant gene can been passed through the generations and consistently causes a single type of cancer. A good example of this type of tumour is retinoblastoma. In other cases a single variant gene can be inherited and result in the evolution of more than one type of cancer, as in the case of multiple endocrine neoplasia type 1 syndrome (MEN-1).

An inherited genetic component is not so obvious in the common tumours, but inherited predispositions for most types of cancer have now been implicated in certain individuals. In some cases there is a clear association between a variant gene and the onset of a common tumour, but this affects only a minority of patients with that tumour. This type of variant gene appears not to be relevant to most individuals with that tumour, as is the case with the variant *BRCA* genes in patients with breast or ovarian cancer. In other cases there are unusual family groups who exhibit a high incidence of cancer and have family members presenting with different tumour types. Although the genes associated with such groupings may be unknown, the involvement of an inherited variant genes is clearly suggested and such genes may well code for DNA repair enzymes which protect the DNA against damage. If these enzymes are not as efficient as in normal individuals then a wide spectrum of damage is possible leading to the occurrence of different cancers in different individuals, but at a higher rate than observed in the general population.

This means that where variant forms of genes are known to be associated with particular cancer, diagnostic tests can be developed to determine whether any members from an affected family group carry the gene and are at risk. Awareness of potential is not just restricted to these familial cancers. There are now screening procedures available for several types of common tumour. Self-examination for breast, skin and testicular cancers are invaluable aids to early detection. If these are carried out routinely they could alert you to early warning signs and, if you find this tedious, then ask your partner to do it for you. Hospital screens are equally useful for breast and cervical cancers. An awareness of any unusual changes needs to be acted upon and therefore knowledge of some of the more common symptoms that are associated with cancer is useful.

Familial Cancers

Retinoblastoma is a rare form of cancer affecting five in 100,000 people. Of these about 40% have a strong inherited component and the remainder are spontaneous. Interestingly most of the individuals from affected families develop tumours in both eyes, whereas in the spontaneous cases tumours tend to be found only in one eye. The gene that appears to be important in this malignancy is an altered form of the *RB-1* gene.

There are a number of other rare tumours originating from neuro-ectodermal tissue which have similar associations with variant forms of genes and it may be that such genes are normally involved in neural development. These include MEN-1 which results in the appearance of tumours at various sites including the anterior pituitary gland, pancreatic islets and the parathyroid gland, MEN-2 predisposes patients to develop tumours in the adrenal gland and thyroid gland. Neurofibromatosis type 1 predisposes patients to a range of tumours including neurofibrosarcomas and gliomas and neurofibromatosis type 2 predisposes patients to meningiomas and neuromas. All of these cases are relatively rare and when they do occur tend to cluster in family groups.

The *BRCA* genes have been shown to exist in variant forms that are associated with breast and ovarian cancer. These are associated with about 5% of breast cancer cases, but where they do occur they are very significant risk factors for breast cancer. They result in breast cancer in women at much younger ages than is normal with breast cancer. Women with these variant *BRCA* genes may develop breast cancer in their 30s and in both breasts. These genes are perhaps more significant than the variant genes in the rare cancers mentioned above because of the very large number of cases of breast cancers that occur each year. Breast tumours that are associated with variant forms of the *BRCA* genes may amount to over 50,000 cases world-wide each year.

Awareness of Potential Risk — the Key to Early Diagnosis and Cure

Where variant forms of genes have been identified as causative factors in cancer occurring in family groups, there is an opportunity for the development of diagnostic tests which will enable the detection of such variant genes in individuals born into those affected families. This type of analysis is very useful because it can identify which individuals are at risk from developing particular cancers. This has two advantages. Firstly it can rule out individuals who may not have inherited the variant form of the gene. In the past all babies born into families affected by retinoblastoma had to be regularly screened for signs of tumour development in their eyes. However, if they can be demonstrated to be free of the variant form of the gene by an analysis of their *RB-1* genes, then not only does this bring relief to the individual and their family that they are not at risk for developing this type of cancer, but it also represents a considerable saving on health service resources as they no longer have to be screened. For those individuals who are found to have the variant form of the gene, the surveillance can be maintained to detect early signs of tumour formation.

This type of screening may not be practical for common tumours where the entire population may be have to be screened, however, where there appears to be a clustering of breast or colon tumours in a family then the relevant genes may be screened to identify whether any (so far) unaffected family members are carrying the variant form of the gene. If they are found to be carriers and therefore at risk they can be monitored regularly for early signs of disease, counselled on life style risk factors that they should pay particular attention to, and may even choose to opt for radical preventative measures.

The ability to detect predisposing genes should be regarded as a positive step as it provides an early warning of possible future problems. If these findings are heeded it could provide an invaluable aid for dealing with a tumour if it did occur, and would help stack the odds in your favour of surviving the cancer.

Screening

Skin cancer may be detected in its early stages by **self-examination**. Most people have moles on their skin and they are of little significance as they are a form of benign tumour. However, if one of these begins to change then it is important to consult a doctor. Changes which may be of importance include altered shape, colour or size. If the mole begins to weep or bleed, then this may also be of concern. Also if a new lump occurs and does not disappear within a few weeks, then this is a matter that should be discussed with your doctor. There are three main types of skin cancer. The basal cell carcinomas and squamous cell carcinomas make up 75% and 20% respectively of all skin cancers. They rarely metastasise and are easily curable. The remaining 5% are the melanomas and they are far more dangerous. However, if detected early enough, even melanomas are curable. The carcinomas tend to occur on parts of the body that are frequently exposed to the sun, such as the head, neck, feet, legs and trunk. However, melanomas can occur anywhere, even on the soles of the feet.

Breast tumours are usually first noticed by a woman as a small lump in her breast. It is important to note that 90% of lumps in the breast are either benign or are cysts (fluid filled swellings), which are not life threatening. Regular self-examination is very important and leaflets describing the best procedures for self-examination are available from a woman's doctor. It is important that a woman is aware of the changes that her breasts go through during the different stages of the menstrual cycle. By following a routine of self-examining she will become familiar with her own breasts and therefore more receptive to noticing any changes. The type of changes which may be of importance include the presence of a lump, thickening of the breast, a change in outline of the breast, dimpling of the skin on the breast, discharge from the nipple, a lump or thickening of the nipple, unusual inversion of the nipple, swelling in the armpit and unusual discomfort or pain in the breast. The importance of early detection is that it provides a much greater chance of a cure. For example, when very small tumours are detected there is a better than 90% cure rate. With tumours of 1cm in diameter the cure rate is still good at 70-80% but the larger the tumour the worse the outlook.

Breast screening programmes have been established in many industrial countries, with the intention of detecting breast tumours at early stages of development. Treatment of such small tumours is often more successful than for more developed cancers. These programmes are generally available to women over 50 because breast cancer is far more common in older women. Also the breast tissue in younger women is more dense and therefore it is often difficult to tell whether a small tumour is present. This involves a clinical examination and mammography. Mammography involves taking an X-ray picture of the breasts to determine whether any small lumps that may be tumours are present. These screens are normally scheduled every three years, but can be carried out at any time if you experience a change in your breasts. In

1992/1993, in the United Kingdom, over a million women responded to an invitation to undergo mammography. Of these, 63,000 were recalled for further investigation, which in turn resulted in the detection of breast cancers in 6,600 women. Although many women were called in for further investigation only 1 in 10 proved to have a tumour so women should not be too unduly concerned about being invited back after the initial screening. It has been estimated that detecting breast tumours at these early stages has produced a 30% reduction in breast cancer deaths.

Testicular cancer appears to be an increasing problem and can develop in men of all ages, though it seems to be most prevalent in males between 15 and 40. The cure rates for this tumour are surprisingly high, but treatment is most effective when the cancer is detected early. Detection of changes associated with the testicles can be achieved by self-examination, maybe once a month. Your local health clinic will probably have leaflets describing the best procedure for this self-examination. Examination is best carried out whilst having a warm bath or shower, to relax the skin of the scrotum, so that changes in the size or shape of a testicle can be detected easily. Cancer rarely develops in both testicles and therefore one can act as a control for the other. Generally speaking men do not experience pain in the testicles after cancer develops. Other warning signs may be a general sensation of heaviness in the scrotum, or an ache in the lower abdomen. In a few cases men may experience swelling of the breasts.

Cervical cancer can be screened for, at a medical centre, by use of the Papanicolaou (Pap) smear and pelvic examination. This procedure has generally been an effective way of early detection and has lead to a major decrease in deaths from cervical cancer during the past 40 years. Although there are clear benefits of Pap smear screening, substantial numbers of women still do not bother to make use of this opportunity. This is particularly the case in older women. The procedure is simple and involves taking a smear of cells from the lining of the cervix. These are then stained and examined microscopically for the presence of abnormal cells. These cells are graded from 1 to 3 depending on how different they appear from normal cells. It has been estimated that if women are screened every five years between the ages of 20 and 64, their risk of developing invasive cancer of the cervix is reduced by 80%. It may be that as a result of the smear test you are recalled to the hospital. This does not necessarily mean that evidence of cancer has been detected, it may simply be that insufficient cells were found on the original smear and another one needs to be carried out.

Cancer Symptoms

It is always important to be aware of changes in your body and how they might be early indicators of the development of cancer. This is not to say that you should turn into a hypochondriac or neurotic, but simply that certain symptoms need to be assessed by an experienced medical practitioner to ensure that any appropriate steps are taken. The following symptoms are those which often accompany some of the most common tumours. They are not, however, always indicators of cancer and may well have other less serious causes.

The most usual symptoms of lung cancer are a cough that does not get better, even after treatment with antibiotics, and the coughing up of blood. These may also be accompanied by a shortness of breath, a pain in the chest and general feelings of weakness and weight loss.

The most common symptom of stomach cancer is persistent indigestion. Other minor symptoms may include a bloated feeling after eating, sometimes followed by nausea or vomiting. Loss of appetite, weight loss, or passing of blood in the faeces may also occur. The vague nature of these symptoms means that early diagnosis is not easy and so an opportunity for early diagnosis may be lost. Usually patients are first diagnosed after they have consulted their doctor for more acute symptoms. In Japan where stomach cancer is common, the first sign of indigestion in a patient is investigated for the possibility of stomach cancer. This means that many more tumours are detected at an early stage than is the case in the West, and hence the survival rates for stomach cancer are better in Japan. The problem with investigating all patients with indigestion is that the symptoms are often the result of indigestion and nothing more serious. However, it is probably wise for anyone over 45 who has persistent indigestion to consult their doctor.

Colon and rectum tumours are curable if detected at early stages. Their most common symptom is blood in the faeces. This should be checked out by a doctor, though the most likely cause will be haemorrhoids. Another symptom which often occurs is diarrhoea or constipation, or both alternatively. If such symptoms persist for more than two weeks then a doctor should be consulted. There are no routinely available diagnostic tests available but it is possible that a test based on the detection of mutant *P53* genes (which is the most common mutation in colon cancer) in the faeces may be a possibility in the future. There are some families with a high incidence of colon tumours and in this case a regular examination and faecal blood test may be useful in individuals over 35.

The early stages of prostate cancer do not normally cause symptoms.

Though a swelling may later be noticed, and pain may be experienced in the passing of urine. The passing of urine may be more frequent than usual and also occur during the night. Prostate cancer is becoming very common and can occur in men from 45 to 80, though the incidence greatly increases from 60 onwards.

The first symptom of liver cancer is a feeling of discomfort in the upper abdomen, possibly caused by the enlargement of the liver. It is also possible that the growing tumour may obstruct the bile duct. In this case there will be a build up of bile in the blood which will result in jaundice. This may also cause the urine to become a dark colour and the faeces to become pale. Swelling of the abdomen due to a build up of fluid may also be present. As mentioned before, symptoms such as these may be the result of causes other than cancer, but it is important that they are referred to the doctor.

Enlargement of lymph nodes is the first sign of lymphoma (Figure 4.1), but it is much more commonly associated with an infection. The lymph nodes are found mainly in the neck, armpits and groin. The enlargement of lymph nodes as a response to an infection usually subsides within a few weeks but, where a lymphoma is present, the swelling does not disappear and in fact grows larger. Some individuals have night sweats, weight loss (more than 10% of body weight) and generalised itching. The form of lymphoma called Hodgkin's disease tends to be localised to a group of lymph nodes. It shows a bimodal distribution in that occurs mainly in individuals between 15-34 and over 50. It responds well to treatment which often results in a cure. The non-Hodgkin's lymphomas may occur at several lymph node groups and also be associated with other tissues. They are more common in older age groups and their response to treatment is variable with the majority of patients eventually dying from their disease.

These summaries of the most common symptoms that occur in the most common cancers indicate that symptoms do occur that can alert the individual to potential problems. In the majority of cases the problems are not related to cancer, but rather to some other usually less dangerous condition. However, by being aware of changes that are happening to your body and consulting with your doctor it is possible to identify cancer at an early stage if it is present and, as Galen pointed out 1,800 years ago, the earlier a tumour is treated the better the chance of a cure.

Chapter 11
RADIATION RISKS

Introduction

There are many kinds of radiation to which the body is exposed during its life. By far the most common is sunlight and it is the ultra-violet component of sunlight that is of concern. In parts of the world where there is a high level of sun and people have lived for hundreds of thousands of years, there has been a gradual adaptation to these conditions which has seen the development of high levels of melanin pigment in the skin. Such individuals are therefore well protected against the dangers of ultra-violet light. The problems occur primarily in white people who have either moved to such sunny climates or travel there for short visits. With a lack of protective melanin in the skin, they are very much more at risk from the damaging effects of sunlight.

Although atomic power and nuclear explosions have taken hold of the public imagination, their effect on cancer incidence appears to have so far been negligible. That is not to say that such sources of ionising radiation are safe. Quite the contrary, they are potentially very dangerous. It is just that humans are not exposed to such radiation to the same extent as they are to sunlight. Public perception of risk is often at odds with reality. Most exposure to ionising radiation occurs as a result of naturally occurring **radon** gas permeating homes in regions with underlying granite rock, or through exposure to X-rays in hospital (Figure 11.1).

Sunlight

One of the greatest increases in cancer incidence during the past 40 years has been associated with the occurrence of skin cancer in white people. In the United Kingdom alone more than 40,000 people are diagnosed with some form of skin cancer every year. There are several types of skin cancer and in many cases they are relatively easy to deal with, but one form called melanoma is particularly aggressive and readily spreads to other parts of the body.

It appears that the primary cause of these malignancies is ultra-violet light, which is present in sunlight. There are two main forms which are

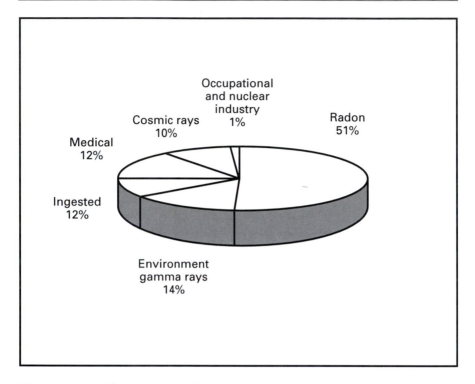

Figure 11.1 Main sources of ionising radiation in the environment.

termed UVA and UVB. There has been a tendency for sunbathing to become a major leisure activity, particularly for white Europeans and Americans during this century. However, the protective pigmentation that is present to varying degrees in black, Asian and even oriental populations is underdeveloped in white skin. This means that there is a much greater risk of the sun burning the skin in white individuals. The effect of **sunburn**, particularly when it occurs in children, is potentially dangerous as the ultra-violet light can cause mutations in DNA, which are of course the initiating steps in transforming a cell into a cancer cell.

This is not to say that sunlight is entirely a bad thing. On the contrary it encourages the formation of Vitamin D, which is important to the body as it helps it assimilate calcium. This is particularly important in Asian and Afro-Caribbean people. Like most things it is a question of balance.

The implications are clear and they are that if you have a white skin, then you should take care not to expose yourself to too much sunlight at any one time. White children and babies may be of particular risk because they have the lowest levels of pigment in their skin and because they have more of their life ahead of them in which cancers can develop. If you wish to develop a suntan, then it will have to be a slow procedure. This involves gradually building up a tan, which is a protective pigmentation of the skin. If you want to spend longer in the sun, then it will be necessary to use a sun block. These creams are claimed to reduce ultra-violet light penetration to the skin, and thus can increase the time that can be safely spent sunbathing. It should also be remembered that the closer one is to the equator, the thinner the atmosphere and the stronger the sunlight.

One of the ways that has been suggested to improve tanning is by using a **sunbed**. This is because ultra-violet light occurs in two forms and sunbeds only give out UVA. Both induce the tanning procedure, but previously only UVB was thought to damage DNA and be linked to causing cancer. The principle of the sunbed is therefore to generate the 'safe' form of ultra-violet light. Although UVB is certainly the main cause of sunburn, several authorities have questioned whether a safe form of ultra-violet light really exists. UVA is more penetrating than UVB, it ages the skin and causes wrinkling, and has recently been linked with skin cancer. It may well be several years before we know whether sunbeds are truly safe forms of tanning. In the meantime it is probably best not to risk it.

The general advice, particularly for white people and especially if they have many moles on their body or have red or fair hair, is to avoid the sun during the hottest part of the day. If you do need to be outside, then cover up. Wearing a hat protects not only bald patches on the top of a man's head, but also protects the tops of the ears and the face. The latter are both common sites for skin cancer. Sunglasses with a UV filter are preferable because the eye is also susceptible to melanoma.

Suntan lotions are useful protective agents and ones with higher **sun protection factors** (SPF) are best. As a general guide, the SPF is a multiplication factor of the time you can safely stay in the sun. Therefore if you normally suffer sunburn after 30 minutes, then a lotion with a SPF of 10 should allow you to stay outside for 300 minutes or 5 hours. This is a rough guide and it is always better to err on the side of caution. The use of lotions with SPF of 15 or above are usually recommended for white individuals and it is important to remember that they wash off during a swim in the sea. Even sweating may cause some of the lotion to become dislodged and regular application of lotion is advisable. As skin cancers often arise on the lips, the use of a lip balm with a SPF of 15 would also be useful. Suntan lotions do seem to protect against sunburn, but some researchers have suggested that they do not protect against melanoma, and instead suggest that genetic factors are more important. Lotions which contain **octyl-dimethyl PAB** as an active ingredient, have also been criticised as potentially carcinogenic in their own right.

Damage to the skin can take many years to cause enough mutations in skin cells so that they form tumours. It is best to avoid constant exposure to bright sunlight and very important that sunburn is prevented.

Ionising Radiation

Ionising radiation includes X-rays and discharges from radioactive chemicals. It is generally assumed that the low levels of X-rays that patients are exposed to at infrequent visits to radiological departments in hospitals do not pose a major risk. They are however kept to a minimum and more frequent sources of exposure, such as X-ray machines in shoe shops, are now banned.

The sources of radiation from chemicals have gradually been identified during the twentieth century as potentially dangerous and these are now either monitored rigorously or prohibited. The luminous dials that were hand painted onto watches resulted in many cases of mouth cancer. This was caused by the painter licking their brushes to obtain a fine point in between dipping them into radioactive paint containing radium and thorium. Workers in industries involved in the extraction of radium from pitch blende developed leukaemias and bone cancers.

There is still a continuing controversy about the risks of developing cancer when living near to a **nuclear power** plant. It has been suggested that the clusters of childhood leukaemias that have been observed in such localities is proof of a link, but this is far from certain. It has been suggested that support for this idea has been provided by the high incidence of leukaemia that occurred in survivors of the **atomic bomb** attacks in Japan during the Second World War. What is often not mentioned though is that the Hiroshima and Nagasaki survivors invariably developed chronic myeloid leukaemia, a disease which is very different to the acute leukaemias that occur in the children near power stations. Furthermore, the Japanese people were subjected to a very large dose of radiation at one time, which is different to the very low dose exposure over a period of time experienced by people living near nuclear installations. One of the problems in understanding this situation is that the numbers of affected children is very low and it is difficult to draw meaningful conclusions at present. However, another possible explanation has recently been suggested. It is possible that men working in the nuclear plant are exposed to relatively high levels of radiation which cause mutations in the germ cells and these are passed on to their children, who are then predisposed to develop leukaemia. This is an interesting idea, but not proven.

Nevertheless the potential dangers of radiation in causing cancer remain real and need to be continually monitored. This is primarily a concern for government regulation as was demonstrated with the recent release of radioactive material from the nuclear power plant at Chernobyl. So great was this release that it resulted in an increased incidence of cancer, particularly thyroid cancer, locally. However, radioactive material was also carried into the atmosphere and winds blew it to north west Europe. There was a possibility that radioactive chemicals may have entered and been concentrated by the food chain. As a result, governments acted by slaughtering, and then destroying the carcasses of, thousands of food animals. Reindeer were destroyed in Finland and sheep in Scotland. In this way the government sought to reduce the risk to the human population.

Another situation which may involve a risk and which the individual could do something about themselves concerns radon gas. It has been noted for some years that in certain areas, particularly where the underlying rock is granite, that ground rocks give off low levels of radon. Radon is a radioactive gas and is generally present in levels which are not thought to constitute a health threat. However, it appears that houses built in these areas may act to concentrate the gas, which can lead to the build up of levels of gas and radiation which are regarded as being above acceptable levels. There is still debate as to whether this is associated with an increased incidence of lung cancer in people who live in the affected houses. It is perhaps a wise precaution if you live in such an area to have your house monitored and if necessary fit a pumping system (these need not be too expensive) to reduce the build up of radon inside the house to acceptable levels. A survey recently carried out in the south-west of England found that 5% of properties had levels of radon above 200 becquerels per cubic metre of air. This may be a particular problem in some modern houses which have double glazing that could concentrate the gas inside the house.

In the United Kingdom, the Department of the Environment have produced a booklet, The Householders Guide to Radon, which provides more advice on how to reduce the levels of radon gas in a house.

Chapter 12
DIET, OBESITY AND CANCER

Introduction

Diet is undoubtedly one of the most important factors which influences the risk of developing of cancer. Table 12.1 summarises the current scientific view on how many cancers could be avoided by adopting healthy diets. Although there is a certain amount of variability in the influence that diet has on the incidence of different cancers, the clear conclusion is that at least one third of all cancers are avoidable by dietary changes. Individuals who have a diet rich in fruits and vegetables with low levels of animal fat, will have a greatly reduced risk of developing cancer.

It is very difficult to design definitive guidelines concerning diet, as diet can vary considerably around the world. There is little point recommending people in Japan to reduce their fat intake, when it is already relatively low. What would be of more import in Japan is that people reduce their salt intake, which is much higher than in the USA or Europe. It is in the latter countries where fat content in the diet is an important factor in cancer incidence. The following sections should therefore be considered in the light of the local situation. The general points made here are applicable to the overall risk of developing cancer and some indication is given as to what is a reasonable intake of the different components of the diet.

One other important fact to note about these recommendations is that a healthy diet as far as cancer risk is concerned, is also a healthy diet

Table 12.1 Cancer deaths avoidable by a healthy diet.

Cancer type	Percentage of cases avoidable
Colon	90
Stomach	90
Breast	50
Pancreas	50
Bladder	20
Lung	20
Oesophagus	20
Prostate	10

for reducing the risks of other health problems, such as heart disease and diabetes. A diet that is protective against, and carries few risks for, developing cancer is basically a good diet, which will contribute to your general well-being.

The conclusions that have been formulated about the impact of diet on cancer incidence are the result of very many studies over the past few decades. These were summarised in a major report, 'Food, Nutrition and the Prevention of Cancer: a Global Perspective', which was published in 1997 by the World Cancer Research Fund in association with the American Institute for Cancer Research, and a more detailed analysis of risk can be found there.

Obesity

The risk of developing cancer appears to increase in overweight individuals. This also parallels the increased incidence of heart disease and diabetes. Increase in weight is often associated with energy-dense diets, particularly in industrialised countries where many individuals lead a sedentary life. It is possible to gain an indication of how healthy one's weight is by referring to **body mass indices** (Figure 12.1). By comparing an individual's height and weight on one of these figures it is possible to judge whether the individual has an appropriate weight for their height, or whether there are underweight, or overweight and by how much.

In recent years there has been a growing movement to suggest that fatness is socially acceptable and that obese people should not be criticised for their weight. This sort of politically correct nonsense does no one any favours. The link between obesity and premature death, from a number of causes, is clear. There is accumulating evidence that obesity increases the risk of developing sarcomas. It is also likely to increase the incidence of breast cancer in post-menopausal women, and possibly cancers of the colon and kidney in both sexes. The risk of breast cancer in women appears to be related to early onset of the menopause, which is in itself related to being overweight or obese. However, there have been several reports published recently which have also indicated that there is an increase in the general cancer incidence in individuals who are designated as overweight though not obese. Although smoking tobacco may limit weight gain, any protective influence this may provide is greatly outweighed by the enormous increase in tobacco-induced tumour formation.

Another cause for concern is rapid growth in childhood and the

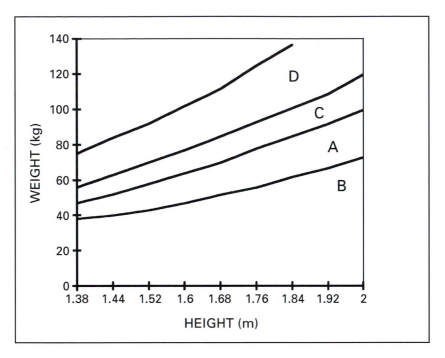

Figure 12.1 Body mass indices. By comparing weight against height an individual can determine how healthy is their weight. (A) The ideal weight to height area. Individuals who have a sedentary occupation should preferably be towards the lower part of this area, whereas those who have a physically active life would be towards the upper part of the area. (B) This area indicates the individual is underweight and possibly malnourished. (C) This area indicates the individual is overweight and therefore at a slightly increased risk of developing cancer. (D) This area indicates that the individual is obese and is clearly at risk of developing cancer.

accompanying overweight during infancy and adolescence. This is because such people tend to remain overweight into later life and may become persistently obese. There is also some evidence that girls who are overweight prior to puberty may be at an increased risk of breast cancer in later life.

The relationship between low body mass and cancer incidence is unclear. This is because smokers are on average thinner than non-smokers, but because of their smoking have a much greater risk of cancer. It is also possible that underweight individuals may be showing symptoms due to a developing cancer which is not at present otherwise apparent.

Related to body weight is the question of **exercise**. There is good evidence that regular physical activity is protective against colon cancer and there is a suggestion that it may also offer some protection against breast and lung cancers. It has been recommended that if a person has a sedentary occupation, then they should take some exercise (eg., brisk walking, cycling, gardening) for one hour each day and undergo strenuous activity (eg., running, swimming, playing tennis, walking up and down stairs) for at least one hour each week.

The relationship of **energy** content of the diet and cancer risk is complicated and links with cancer risk may be actually be due to other factors that are related to energy balance itself. However, experiments on many animal species have now demonstrated that diets which have consistently low levels of calories, from whatever source, delay the onset of cancer.

Food Constituents

Carbohydrate

Carbohydrates comprise sugars, starches and non-starch polysaccharides (**NSP**), and are the main source of dietary energy in most regions of the world. NSP is important as it provides much of the **dietary fibre**. Starch is found in cereals, pulses and roots and are usually cooked in some way before consumption. In developing countries the majority of total dietary energy comes from starchy foods, but in industrial countries less than half comes from starch and an increasing proportion of this is provided by refined sugar. NSP is also an important component of fruit and vegetables.

Diets which are high in starch have been associated with a low risk of stomach cancer but a low protective effect against colon cancer. This confusing observation may be explained by the fact that high carbohydrate diets frequently also contain high levels of salt (see below) and low levels of fruit and vegetables, both of which are associated with stomach cancer. Therefore, the important effect of starch may be more to do with protecting against colon cancer. However, if the diet is rich in refined sugar rather than NSP then there is a low risk of colon cancer. This may occur because a high level of starch (particularly NSP) in the diet could reduce the transit time in the colon, where a diet rich in sugar would slow transit time. High NSP diets also appear to have a low protective effect against cancers of the breast and pancreas. Refined cereal foods such as white bread and sugars also are associated with a slight risk of oesophagus cancer, though the salt content of such processed foods may be an important factor in this case.

Fats

Fats may be classified as saturated, monounsaturated or polyunsaturated and may be of plant or animal origin. In addition to these is **cholesterol**, which although not technically a fat has related effects, and is of animal origin. Fats are present in most foods, though the fat content of animal foods is generally greater than that of plants. Fats are often used during cooking and vegetable oils, butter, lard and margarine are composed almost entirely of fats. Fats are a necessary part of a healthy diet and it has been suggested that between 15–30% of total dietary energy intake should be provided by fat, of which less than one third should be saturated fat.

A diet with a high total fat content is associated with a low risk of cancers of the breast, colon, lung, prostate and rectum. Diets which have a high saturated fat content, particularly from animal foods, also have in addition an associated low risk of sarcomas. Diets high in cholesterol are associated with a low risk of lung and pancreatic cancers. Of course diets high in fat greatly increase the risk of obesity and therefore are an indirect risk factor for all types of cancer.

In the developing countries most fat is of plant origin and fat intake contributes only a minority (20%) of total dietary energy. In industrialised countries fat intake can provide over a third of dietary energy and the majority is of animal origin, mainly saturated fat.

Fat content in fruit and vegetables is generally low. Cereals may contain from 0.5–8% of fat, and nuts have an even higher content. It should be remembered that baked products such as biscuits, cakes and pie pastry frequently contain high levels of fat.

Protein

There is no convincing evidence that dietary protein, either from plant or animal sources, has any effect, either protective or as a risk factor, in cancer development.

Salt

Salt is an essential component of the diet as it is required for the regulation of body fluids. It has been estimated that the average adult human needs a daily intake of 0.5g although the usual intake for many people is anything up to 14g. In most industrialised countries the daily salt intake is 8–10g, but in south east Asia can be 10–12 and even up to 14g in parts of China. There is clear evidence that high levels of salt in the diet are associated with developing stomach cancer and naso-pharyngeal cancer.

The salt contents of naturally occurring foods are generally low, and salt in diets is mainly from salt added to preserve foods, in their manufacture, in their cooking and at the table. Salt is still extensively used to preserve foods, as it has been for thousands of years. There can be as much as 5g salt in a 100g portion of bacon or sausage, and in salted fish it can be as high as 10g per 100g portion. Savoury snacks such as crisps (potato chips) and peanuts can contain up to 5g per 100g portion. However, the salt content can also be high in other foodstuffs that one might not immediately regard as salty. Bread and breakfast cereals, for example, can contain up to 4g per 100g portion. The addition of salt to manufactured foods is common, so that even relatively bland foods such as butter may contain substantial quantities of salt.

Salted foods are very common in the diets of people from south-east Asia. Not only are salted fish and pickled vegetables (which contain high levels of salt) frequently eaten but there are also high salt levels in soy sauce and miso (a soybean paste). It has been estimated that such condiments contribute half the salt intake in Japan. Interestingly, this part of the world has high levels of stomach cancer. In Europe the highest levels of stomach cancer are found in Eastern Europe, where again pickled vegetables such as sauerkraut are commonly eaten.

Interestingly the latter part of the twentieth century has seen a significant decline in the incidence of stomach cancer in industrialised countries. It has been suggested that this has been the result of a growth in the use of refrigeration both domestically and by the food manufacturers. Food that can be frozen or kept cool does not need salt preservation and such foods now form a smaller part of the diet in those countries.

Vitamins and Trace Elements

There are many **vitamins** present in foods and a number have been investigated for their association with cancer. Generally speaking a number of protective effects have been identified for diets with high vitamin contents. Vitamin C has a moderate protective influence against stomach cancer, as have carotenoids against lung cancer. Low level protective effects have also been concluded for vitamin C against cancers of the cervix, lung, mouth and throat, oesophagus and pancreas. Similar protective effects have been identified by carotenoids against cancers of the breast, cervix, colon, oesophagus, rectum and stomach. Vitamin E also appears to have a low protective effect against cervical and lung cancers.

The carotenoids are found primarily in vegetables such as carrots, pumpkins, cantaloupe, squash, spinach and fruits such as apricots and mangoes. Vitamin C is present in vegetables such as broccoli, cabbage, peppers and tomatoes and in citrus fruit, mangoes and strawberries. Vitamin E is present in vegetable oil, margarine, nuts and wheat germ. A well balanced diet containing fruit and vegetables should provide a high level of these vitamins. In these circumstances supplementing the diet with vitamin tablets has no further protective effect against cancer. However, in poor quality diets, vitamin supplements may be useful in raising daily intakes to the high levels found in well balanced diets.

It has been shown that both carotenoid and vitamin C levels in the blood of smokers are only 75% of those observed in non-smokers, even though their dietary intakes were similar. This might provide another mechanism by which tobacco usage can increase cancer risk.

So far only two **trace elements** have been associated with cancer risk. One of these is selenium which if present at high levels in the diet is associated with a low protective effect against lung cancer. Selenium is found in cereals, meat and fish and is particularly high in Brazil nuts. The other element is iodine and has been implicated in the development of thyroid tumours. Iodine is an essential component of the hormone thyroxin which is manufactured in the thyroid gland in the neck. The best source of iodine is from seafood. It appears that iodine deficiency in the diet is associated with a moderate risk of thyroid cancer, but too much iodine in the diet is associated with a slight risk of thyroid cancer. Again this is a question of balance.

Food Additives

Most of the colouring and flavouring agents that are added to processed foods, to make them more appealing and to extend their shelf-life, appear to have no influence on cancer incidence. These include the governmentally regulated use of E-number compounds and sweeteners such as saccharin. The issue of salt is dealt with separately above. It appears that when used in approved amounts, food additives do not constitute a cancer risk.

Food Contaminants

The most important type of food contamination with regards to cancer concerns the growth of **moulds**. Moulds are particularly problematic in tropical countries which have poor storage facilities. The hot and humid conditions under which food is stored encourages mould growth on the food and as moulds contains a number of chemicals, called mycotoxins, these can then contaminate the food. Certain Aspergillus species of mould are known to produce a particularly toxic chemical called aflatoxin. This is of concern because consumption of aflatoxin contaminated food is associated with an increased risk of liver cancer. The liver cells are involved in metabolising such chemicals and therefore are most susceptible to their mutagenic effects. This situation may be exacerbated when individual have also been infected by hepatitis B virus.

World-wide it has been estimated that a quarter of the world's food crops are annually contaminated by mycotoxins though this contamination is found more in developing countries where food storage conditions are poor. The main crops that are involved include beans, cereals (barley, maize, oats, rice and wheat) and nuts. Although foodstuffs of this type may be imported into industrialised countries, the greatest health risk is in Africa and south-east Asia, which mirrors the high levels of liver cancer in these regions.

There are other mycotoxins that have been reported to contaminate food crops in industrialised countries but their levels are not thought to constitute a major problem. These include fumonisin mycotoxins in maize from Europe and North America, and orchratoxin A in barley in the UK.

An additional problem may also arise in regions where contaminated feed is given to cattle as it can result in secretion of the mycotoxins in their milk. This again can pose a threat through human consumption of the contaminated milk.

Food Types

Fruit and Vegetables

There is very good evidence that diets which are high in fruits and vegetables offer better protection against many types of cancer than diets containing meat. What is not so clear is whether this is because these diets are intrinsically healthier or whether it is because they contain less foods which are associated with an increased cancer risk.

Certainly fruit and vegetable based diets are lower in energy and may therefore reduce the risk of obesity. Therefore a diet may be equally healthy which contains animal products, as long as the overall nutritional contribution is the same as that of a vegetarian diet.

Generally plant foods do not increase cancer risk and their effect is essentially protective. Both fruit and vegetables provide a high protective effect against cancers of the mouth and throat, oesophagus, lung and stomach, and vegetables show a similar effect against colon and rectal cancers. Both fruit and vegetables have moderate protective effects against cancers of the bladder, breast, and pancreas. They also provide low protective effects against cancers of the cervix, ovary, thyroid and against sarcomas. Vegetables also provide a low protective effect against cancers of the kidney, liver and prostate.

These foods are usually low in energy, and good sources of NSP and vitamins. Vegetables appear to be more effective than fruits in protecting against cancer, but this may reflect the fact that they are generally consumed in greater quantities.

The vegetables referred to here include asparagus, avocados, beetroot, broccoli, cucumbers, courgettes, green leafed vegetables, mushrooms, onions, parsley, peppers, sprouts and tomatoes. Vegetables such as beans, cassava, lentils, nuts, peas and potatoes have not so far been shown to have an influence on cancer risk.

Studies of populations in a number of countries who follow a primarily vegetarian diet and do not drink alcohol or smoke, have demonstrated that they have a significantly lower cancer incidence than the general population. Breast cancer may be less common in vegetarian women because they have been shown to have lower levels of hormones, (androgens, prolactins, oestrogens) in their blood. Prostate cancer in vegetarians may similarly be less common because of reduced blood plasma levels of steroid hormones.

There is no evidence at present to suggest that the use of herbs or spices, including garlic, has any effect on cancer risk.

Animal Products

Generally speaking **meat** and **fish** offer no protective effect against cancer and what effects have been noted are to increase cancer risk. Red meat from domesticated animals (beef, lamb, pork) appears to contain more fat than undomesticated animals. Diets which are high in domesticated red meat show a high risk of cancer of the colon and rectum and moderate risks of cancers of the breast, kidney, pancreas and prostate.

The influence of meats such as venison and rabbit is unclear. There is also no clear evidence for either a protective effect or a cancer risk from eating poultry or fish.

Diets which have a high content of eggs are associated with a slight risk of colon and rectal cancers. Also diets with a high dairy product (milk, cream, cheese) also appear to be associated with a low risk of kidney and prostate cancers.

Tea and Coffee

A number of studies have investigated potential risks associated with a variety of hot drinks. Generally no relationship has been found between the drinking of such beverages and the incidence of any cancer. A low protective effective against stomach cancer has been claimed for the drinking of green tea in populations that have a high incidence of this disease.

A low risk of developing cancers of the mouth and oesophagus has also been suggested by the habitual drinking of very hot drinks of any kind. Presumably this is because this behaviour tends to damage the lining of the mouth and oesophagus, which in turn stimulates cells to proliferate during which they may acquire mutations that contribute to their malignant transformation.

Food Processing

It is difficult to assess the relevance of curing or smoking meat and fish, as there are other considerations, such as salt content, which can affect a judgement of how dangerous these procedures are. Cured meats are of course the main dietary source of nitrates, nitrites and nitrosamines. Nitrites are found in lesser amounts in processed cereals and baked products. The concern has been that nitrates and nitrites may be converted into nitrosamines in the stomach, because nitrosoamines are known carcinogens. However, this does not appear to be a significant problem, and it seems that nitrosamines that are already present in cured meats may be of more concern. In industrialised countries there has been a steady decline in the amounts of nitrates, nitrites and nitrosamines in cured meats in recent decades, primarily due to government legislation.

There remains a low risk that cured meats increase the incidence of cancers of the colon and rectum. Smoked meat and fish do not appear to present a problem unless they are regularly included in the diet. In countries where this happens, such as certain East European countries and Iceland, this has been linked to a high incidence of stomach cancer.

There is little evidence that cooking food can affect its cancer risk, however, it is known that cooking at high temperatures particularly with grilling or barbecuing can convert fat in the food into carcinogenic polycyclic aromatic hydrocarbons. These are similar to some of the carcinogenic compounds found in tobacco smoke. It has been suggested that the surfaces of char-grilled meats and fish contain such carcinogenic chemicals whereas similar food that has been fried do not. Overall if a diet contains a regular intake of such cooked meats and fish then there appears to be a low increased risk of cancers of the stomach, colon and rectum.

A Guide to Healthy Eating

Balanced meals

To eat in a healthy manner does not mean to diet, it means to eat in a balanced way. All of the major constituents of food need to be included, but in the right proportions. The basic principles to **healthy eating** are: (a) to use foods that are low in fats, such as fruit and vegetables, starchy foods (pasta, potatoes, rice), and pulses (beans, lentils); (b) to increase dietary fibre by using wholemeal products and eating the whole of fruits and vegetables with edible skins; and (c) reduce salt intake. In trying to adopt a healthier diet it is probably best to gradually change a few things at a time, rather than making a radical change. Not only is this easier, but it is likely to have a more lasting impact on one's eating habits. Of course, regular exercise is also important so that body weight remains within recommended limits (Figure 12.1). This can be attained by ensuring energy taken into the body in the form of calories in food, is used in **exercise**. Weight loss will occur when the amount of energy used in exercising is greater that the calorie content of food eaten.

Many people are confused by what is meant by a high fat diet. In essence, the general recommendation is that less than 30% of all calories consumed should come from fat. In industrial countries the average figure is often 40% or more. Processed foods are frequently a source of excess fat, but it is possible to judge how much fat they are likely to contribute to the diet by checking the calorie and fat contents stated on the packaging. As a general rule, for every 100 calories there should be no more than 3.3g of fat present, which means that 30% of the calories are being provided by fat. If however there were 4.4g of fat, then this would represent 40% of calories, and 5.5g would represent 50% of calories.

Snacks

A useful tip for maintaining a balanced diet is to control impulse eating. People often eat when they are not really hungry, but rather do so when they are bored or under stress or as part of a social event. **Snacks** such as cakes, biscuits, chocolate bars and crisps all contain fat and are often high in sugar or salt, but have a low vitamin content. In other words they represent poor nutritional value. Pastry covered snacks, such as pies, sausage rolls and even vegetable samosas, also have a high fat content.

It is not necessary to stop eating these types of food completely, but they should be regarded as an occasional treat and not eaten every day. In fact, if they are gradually replaced with more healthy snacks, then the craving for them may even disappear all together. Although snacks might offer a useful source of additional nutrients it is probably not a good idea to use them to replace regular meals.

There are of course many healthy snack foods. Low fat yoghurts, fruit and vegetables all offer much healthier alternatives. These need not be boring as some people think. Carrot and celery sticks can be enjoyed with a variety of low fat dips. Whole grain crackers or rye crispbreads can be spread with cottage cheese. Salad sandwiches can be made with pitta bread. There are many alternatives to cream cakes and bags of crisps. It should also be remembered that there seems to be no upper limit as to how much fruit and vegetables can be eaten in a day.

Chapter 13
TOBACCO AND ALCOHOL

Smoking — the Single Biggest Cause of Cancer

Do you smoke **cigarettes**? If the answer is 'yes', don't be a wimp and skip this chapter. Read and think about it. There are a number of factors described in this book that reduce the risk of cancer, but it may be true to say that the five most important of these are:

1. Don't smoke
2. Be born with good quality genes
3. Don't smoke
4. Have a healthy diet
5. Don't smoke

I hope the reader gets the message. In other chapters a number of risk factors are presented where there is a moderate or low probability of a link between a risk factor and certain tumour types. In the case of tobacco there is absolutely no doubt about the causal link. We know that there at least 40 chemicals present in **tobacco** smoke which are carcinogenic and damage DNA. We know that all lung cancer cells contain mutated genes, usually *P53* or *RAS* genes. We know that mutations in these genes will transform a normal cell into a cancer cell. There is a clear cause and effect, right the way down the line. The unethical behaviour of the tobacco companies over the past decades in pretending that their product was not dangerous has been an international scandal. Unfortunately even today governments are still unwilling to tackle this issue head on. In Europe we see, on one hand, the enormous costs to the health services in having to treat tobacco related disease and, on the other hand, the official subsidising of tobacco farmers.

If you smoke then you might be interested to learn that your risk of developing lung cancer has been estimated by different authorities as being between 1 in 3, and 1 in 6. This may seem to be an enormous risk, but it is in fact a lower risk than it would have been, because the apparent risk has been reduced by the fact that 1 in 2 smokers die at an earlier age from heart disease. The message is simple, if you smoke cigarettes then you have a 90% chance of dying prematurely (that's 20 to 30 years prematurely) through either heart disease or cancer.

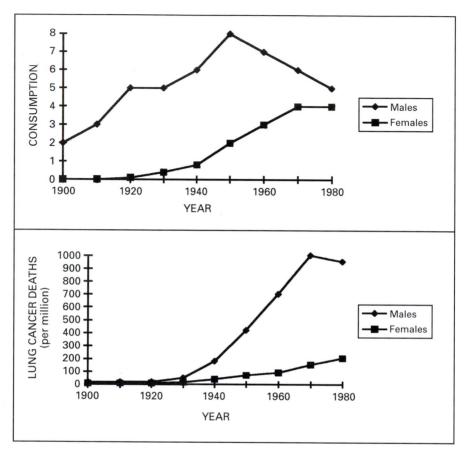

Figure 13.1 Cigarette smoking and lung cancer deaths in men and women during the twentieth century.

Before the twentieth century, lung cancer was a rare disease and it is interesting how the patterns of lung cancer incidence during the twentieth century have mirrored social trends in tobacco usage. The peak of cigarette smoking in industrialised countries was in the 1940s and 1950s, which was reflected in a peak of lung cancer incidence in those same populations in the 1960s and 1970s (Figure 13.1). There is a delay in lung cancer occurring in smokers of at least 15 years. This is important to bear in mind when considering whether to smoke. It means that the longer you smoke and the more you smoke, the greater your chances of developing the disease. However, as damage to cells caused by the carcinogenic chemicals in tobacco may take many years

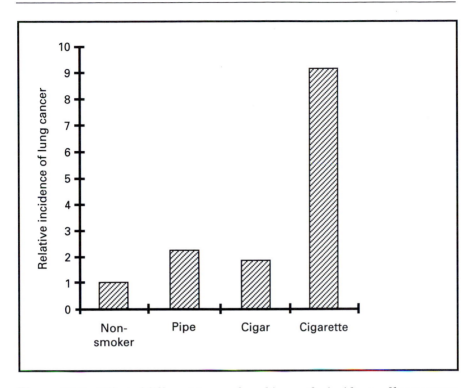

Figure 13.2   Effect of different types of smoking on the incidence of lung cancer in men aged 45–55.

to become manifest as a tumour, you cannot just stop smoking and assume you are free of the consequences of using tobacco. Once you have stopped smoking it may be 15 years before you can feel confident that you will not develop lung cancer.

The risks of developing cancer after using tobacco appear to vary depending on the type of smoking. Figure 13.2 shows the relative risk of smokers developing lung cancer compared to non-smokers. Pipe and cigar smokers, who tend not to inhale much of the tobacco smoke, have double the risk of developing lung cancer than a non-smoker. However, cigarette smokers are almost ten times as likely of developing lung cancer. The results shown in Figure 13.1 refer to men in the 45–55 age range, but the differences would be very much greater in men (or women) at older ages.

The numbers of cigarettes smoked are also an important factor. The more cigarettes smoked, the greater the chance of developing lung cancer. This is demonstrated clearly by Figure 13.3 which shows the

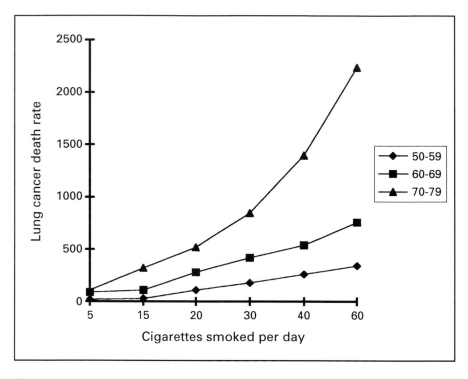

Figure 13.3 Death rates due to lung cancer in relation to numbers of cigarettes smoked per day. Rates are shown for men aged 50–59, 50–69 and 70–79.

correlation between numbers of cigarettes smoked each day and the death rate from lung cancer in men at different ages. The fact that this observation is not due to other regional factors is shown by studies in different countries throughout the world, where the rates of cigarette consumption may vary considerably. Figure 13.4 shows a clear relationship between cigarette consumption and the death from lung cancer in countries from Europe and North America. It is also the case that the earlier a person begins smoking, the greater the risk of them developing lung cancer by middle to old age.

It is probably the case that cigarettes which have a lower tar content carry a lower risk of causing cancer. This is an idea that has been widely advertised by tobacco companies as a way of increasing sales. Unfortunately this has led to the mistaken belief by many smokers that such cigarettes are safe to smoke. Nothing could be further from the truth. It has become clear in recent years that smokers of low tar cigarettes inhale smoke more deeply into their lungs, and that to keep up their

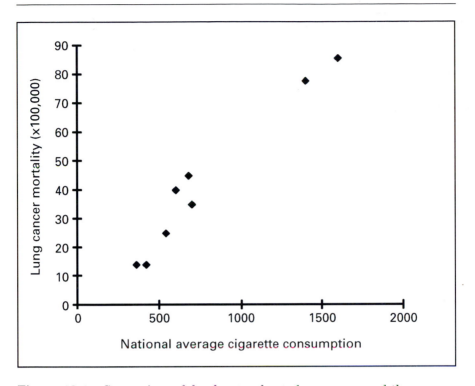

Figure 13.4 Comparison of death rates due to lung cancer and the average cigarette consumption in European and North American countries.

nicotine intake often smoke more cigarettes. This has had the effect of greatly increasing the incidence of adenocarcinomas of the lung. These are a type of cancer that arise deep within the lungs and used to represent a minor form of lung cancer, but today they are amongst the most common type. In the United Kingdom alone 10,000 new cases of adenocarcinoma of the lung are diagnosed each year, and invariably most of those patients die within a few years of diagnosis.

If these facts about the cancer-causing effects of tobacco were not bad enough, it has also been found that cellulose acetate, which is a component of cigarette filters, has been found to cause cancer in animals. It is therefore almost certainly carcinogenic in humans.

The message is stark and quite simple. Don't start smoking in the first place and if you already smoke then stop now. Not tomorrow, now. It is astonishing that some people appear to be so concerned about tiny increases in leukaemia cases near nuclear power stations or very small numbers of variant CJD cases in people who have eaten beef, and yet they show little interest in the hundreds of thousands of people who die each year because of the effects of tobacco. Of course the reason for this is quite clear. It is to do with drug addiction. Tobacco is a complex mixture of chemicals, some of which are carcinogenic. One of the chemicals, however, is an addictive drug and it is called nicotine. Nicotine does not appear to be carcinogenic, but its presence in tobacco means that it produces a dependancy on tobacco usage. As the smoker becomes addicted to using tobacco products, they are exposed to more and more carcinogens and thus the seeds for developing cancer are sown.

The major problem facing society is therefore to find a way of breaking this dependancy on nicotine. There appear to be three main reasons why people initially take up smoking. Some individuals may be subjected to a high stress level because of their job. In such circumstances nicotine can appear helpful in reducing stress levels, but the dependancy that then develops means that the smoker becomes more and more addicted to the tobacco that contains it. To understand how powerful the hold that nicotine may have, one need only consider how many doctors smoke. Here we have a group of professionals who are fully conversant with the risks posed by tobacco and yet many of them still use it. One might ask how can we expect the average person to stop smoking when there are so many health professionals who are well aware of the risks, yet still smoke.

Other individuals smoke because of a lack of self-esteem and confidence in dealing with social situations. Individuals who become nervous when mixing with others find that nicotine has a calming effect. Psychological studies have found this to be an important reason why teenage girls begin smoking. Finally there is a link between smoking and weight loss, and some people clearly use this as a justification for continuing their addiction. It need hardly be pointed out that any potential beneficial effects of losing weight are greatly outweighed by the highly damaging effects that smoking will have to a person's health.

So if you have become addicted to nicotine because of your highly stressful occupation, or because you are socially insecure, or because you want to lose weight, what can you do? The important thing is to stop smoking now. However, some people are particularly susceptible to the addictive effects of nicotine and therefore find it difficult to stop using tobacco. It may be that such individuals might be able to gradually wean themselves off the nicotine addiction by using nicotine patches or nicotine gum. It has been estimated that the success of smokers giving up using will power alone is about 2%, whereas those who have also used one of nicotine substitute aids have a 7% chance. These are therefore useful, but have proved ineffective for numerous users. This may be because the nicotine addiction becomes associated in the minds of the addict with the process of smoking itself. If people are not convinced of this, then they should think of Pavlov's dogs. He demonstrated that he could condition dogs to salivate when they heard a bell, even though no food was present. It may be the same with nicotine addiction and smoking. Recently new ways of delivering nicotine, the nicotine inhalers which resemble cigarettes, have been introduced and time will tell how effective they are.

Interestingly, recent research has indicated that addiction might be due to variant forms of a gene called *CYP2A6*, which codes for a protein that breaksdown nicotine in the bloodstream and brain. The variant form of this gene appears to be less good at breaking down the nicotine and therefore people with it, tend to be less able to tolerate nicotine in their body. As a result they are less likey to become smokers or, if they are, find it easier to give up. The variant form of the gene is twice as likely to occur in non-smokers than in smokers. This may explain why nicotine patches appear to be ineffective in some people, but also holds out the possibility of developing a drug in the future that could inhibit this protein and reduce the craving for cigarettes by smokers.

Probably the best ways of tackling this problem lies in the hands of the government. Financial support for smoking clinics should be provided, as people attending such clinics have a 20% success rate of kicking the habit. The importance of psychology in helping to kick an addiction should not be underestimated. Government support for subsidising nicotine substitutes would also be helpful and, bearing in mind how much smoking-related illness costs the health service, would be a very cost effective measure. Tobacco certainly needs to be taxed out of existence. It needs to be made so expensive that it forces nicotine addicts to reduce their consumption and all aid for tobacco producers must be stopped and their profits taxed at even higher levels. Government also needs to instigate greater awareness of the dangers of using tobacco in schools, so as to reduce the numbers of children and young people becoming addicted in the first place. Remember — cancer cures smoking.

There has been an increased public awareness of the dangers of tobacco in recent years, which has resulted in fewer men smoking. This in turn has led to a decrease in the incidence of lung cancer in men. In England and Wales the numbers of lung cancer cases has fallen by 25% in men. Unfortunately, smoking amongst women has increased during the same period of time and has resulted in a 12% increase in lung cancer.

Recently there have been a number of calls for the legalisation of other materials that contain addictive drugs such as marijuana. This suggestion must be regarded as worrying when one considers the health and social problems that have been caused by nicotine in tobacco products. There are no data at present about the possible carcinogenic effects of long term smoking of marihuana, but the risk must be high that marijuana smoke will also contain carcinogens. Research into this possibility must be a priority before even considering the legalisation of its use.

Passive Smoking — How to Kill your Friends and Relatives

What is particularly worrying about cigarette smoking is the effect that it has on other people, who may be non-smokers. Clear links have been established between exposure to tobacco smoke in the air and an increased incidence of lung cancer. In the open air, this may not pose a risk, but in enclosed conditions with little or no ventilation this is potentially dangerous. This means that a non-smoking partner at home, or a worker in a bar or restaurant, will be subjected to a daily intake of tobacco smoke. Over a period of time this could be the equivalent of them smoking one or more cigarettes a day. Studies of non-smoking wives of Japanese men have demonstrated that the wives of smokers are at greater risk of developing lung cancer than those of non-smoking men. A similar study in the United Kingdom recently found that the risk of a non-smoker developing lung cancer, if they live with a smoker, is increased by 25%. It is reassuring that many workplaces in the United Kingdom and North America now ban their employees from smoking on site. Though it is perhaps regrettable that this has only occurred because of the threat of legal action against those companies by their non-smoking employees.

The most worrying aspect of **passive smoking** is to do with children. Children have little say in their lives and depend on their parents to provide them with a safe environment. It is therefore very sad that a parent's addiction to nicotine can place the child under a serious health risk. Studies have been carried out which have demonstrated that children have more health problems, particularly respiratory difficulties, when one or both of their parents smoke. Certainly cot deaths of new born infants are much more likely to occur if one or both parents smoke. It is astonishing that a quarter of cot death babies have as much nicotine in their bodies as habitual smokers.

It is too early to determine whether the children of smokers also carry an increased risk of cancer in later life, but the chances that this will prove to be the case must be high. Recent evidence has shown that new born babies have been found to have genetic mutations in their white blood cells if the mother smoked when pregnant, or if the father smoked while a non-smoking wife was pregnant. It should be remembered that one of the most common forms of malignant disease in children is leukaemia — cancer of the white blood cells.

It is also the case that, as parents tend to be role models for their children, the children of smokers tend to become smokers themselves thus perpetuating the problem.

Other Forms of Tobacco Usage

I have heard some people say that because they don't smoke and only **chew tobacco** or take **snuff** then this is not likely to cause cancer. Big mistake. I have some bad news for them, because tobacco when taken into the body, in any form, will greatly increase the risk of cancer developing. Chewing tobacco carries a very high risk of causing cancer of the mouth. Snuff is not used so much these days and therefore is not regarded as so much of a problem, but it should clearly be avoided as it has been shown to greatly increase the risk of nasal cancer.

Alcohol

Another addictive drug which has a widespread use in our society is **alcohol**. There is now convincing evidence that alcohol consumption increases the risk of developing cancer. Furthermore, the risk appears to be proportional to the amount of alcohol consumed. So there is no safe limit for alcohol consumption; the more you drink, the greater your chance of developing cancer.

From a world-wide perspective, particularly in Asia, liver cancer primarily appears to be associated with aflatoxin contamination of foods and infection by the hepatitis B virus (see Chapters 5 and 12). However, in the industrialised countries of the Western world, liver cancer often develops from cirrhosis of the liver which has resulted from heavy consumption of alcohol. The organs which seem to be particularly at risk from alcohol are generally those which come into contact with alcohol when it first enters the body, that is the lining of the mouth and digestive tract (though not the stomach) and the liver, where the alcohol is metabolised. Interestingly, a link has also emerged between alcohol consumption and breast cancer. This is important because the link has been observed at relatively low levels of consumption. Women who drink between two to six units of alcohol per day increase their risk of developing breast cancer by 40%. Other tumours that are possibly associated with alcohol consumption are cancers of the colon, rectum and lung. The links between alcohol and the above types of cancer are even more pronounced if the individual also smokes.

It has been estimated that alcohol accounts for 3% of cancer deaths in the USA and this is independent of the form in which the alcohol is drunk. However, it has also been suggested that the majority of these alcohol related deaths would not have occurred had it not been for the synergistic presence of smoking by the affected individuals.

One of the apparent benefits of alcohol consumption, which has attracted much media attention in recent years, is the apparent protective effect that low level consumption of alcohol has for heart disease. This raises the question of how you balance these protective effects of alcohol against its potential cancer risk. In this respect the beneficial effects of protecting against heart disease, which are provided by consuming one unit of alcohol a day, may well outweigh any potential cancer causing risk at this level of consumption. However, with higher levels of alcohol consumption the deleterious effects caused by increasing cancer incidence soon outweigh any benefits to the health of the heart, and in fact at these higher levels of consumption the benefits for heart disease no longer exist.

Alcohol has no dietary value, it is an addictive drug and although it may not cause cancer as a result of a direct action, it certainly appears to function as a co-carcinogen.

Chapter 14
VIRUSES AND CANCER

Introduction

There is now clear evidence that infections by particular types of virus can contribute to the development of cancer. These viruses do not by themselves appear to cause cancer, but rather act as important contributory factors. In other words they can provide an influence, in combination with other factors, resulting in the malignant transformation of cells. What these other influences may be are discussed below.

The fact that viruses are important cancer-causing agents means that their effect might be mitigated by vaccinating people against those viruses. Such vaccinations would of course only have an effect in reducing the incidence of the tumours which are specifically associated with a particular type of virus. In other words, vaccination against hepatitis B virus may be very effective in reducing the incidence of liver cancer, but it would have no effect on the incidence of breast or lung cancer.

Liver Cancer and Hepatitis B Virus (HBV) Infection

Liver cancer is one of the most common cancers world-wide, though its incidence in industrialised countries is much lower than in Third World countries. A few years ago it was established that in the developing countries the incidence of liver cancer correlated with infection by the HBV. More recently another virus, hepatitis C has been discovered and also implicated as a cancer causing agent. This virus may be responsible for many of the cases of liver cancer where HBV was not present.

Infection by HBV can produce a wide spectrum of clinical symptoms. These can vary from asymptomatic transient increases in liver function to severe acute hepatitis and resulting liver failure. HBV infection can also become chronic and produce a range of changes from chronic low level hepatitis through to very damaging cirrhosis.

Studies of patients in Southeast Asia and in tropical Africa have provided clear links with HPV infection and liver cancer. This has lead to major programmes of vaccination against the virus, particularly in Southeast Asia. Interestingly there is evidence that the incidence of liver cancer in these areas may now be showing a decline.

Cervical Cancer and Human Papilloma Virus (HPV) Infection

It is now well established that at least two types of papilloma virus are associated with cervical cancer. This is an interesting virus because it appears to have little effect on men. There are a small number of cases implicating these papilloma viruses in cancer of the penis, but generally speaking they have their cancer causing effect primarily in women. It is not clear why this should be, but it is probably associated with the different hormones circulating in the female body. It has recently been reported that there are variant forms of the *P53* gene which occur in some women which make them particularly sensitive to the effects of the papilloma virus proteins.

There has been a great deal of effort in recent years to develop effective vaccines against papilloma viruses as a way of reducing the risks of developing cervical cancer. Hopefully these vaccines will provide women with protection against this virus and cervical cancer in the not too distant future.

Lymphomas, Head and Neck Tumours and Epstein-Barr Virus (EBV) Infection

Epstein-Barr virus (EBV) was first identified from studies of a childhood lymphoma, called Burkitt's lymphoma, which affected children in Africa. It is in fact a ubiquitous virus and it has been estimated that up to 90% of all people may be infected with it at some stage of their life. Generally infection appears before adulthood and in the developing countries this is often in early childhood. EBV infects the epithelial cells, particularly in the throat, and also a proportion of white blood cells.

The effect that EBV has on the body varies widely from individual to individual. In some cases, EBV results in a stimulation of the proliferation of white blood cells which is known as glandular fever. This is a benign condition which is not malignant. In a small number of cases, however, infection of the epithelial cells or white blood cells can result in the development of nasopharyngeal cancer or lymphoma, respectively.

EBV can produce many different types of lymphoma. It has been linked to almost all cases of Burkitt's lymphoma in Africa, though to only a minority of such cases in industrialised countries. Infection by EBV has also been linked to 40% of Hodgkin's disease cases. In Africa it has been suggested that malaria may be a co-factor, possibly because of its effects in suppressing the immune system. This is an interesting possibility because it has also been noted that in patients that are immunosuppressed (for example, heart transplant patients) approximately 30% develop EBV-positive lymphomas. There is also an increasing incidence of such EBV positive lymphomas in AIDS patients.

The role that EBV plays in the development of malignant disease is not clear, particularly as many people in the population are now known to carry this virus without any apparent ill-health. The fact that it appears to contribute to different health problems, or to no health problems at all, in humans indicates that its influence in the malignant transformation of cells is not strong and that it is dependant on other factors.

Other Viruses and Cancer

There is little evidence for the involvement of other viruses in human cancer. The viruses mentioned above are all DNA viruses. That is their genes are composed of DNA, just like human cells. There is another class of virus, however, that are called retroviruses. The genes of these viruses are composed of RNA and to replicate these viruses must insert their RNA genes into a host cell, where they are converted into DNA and then function as genes. HIV is an example of such a retrovirus. In animals, retroviruses are very common causes of cancer. They influence a cell to become malignant either by the insertion of the DNA copy of their genes into one of the cell's chromosomes adjacent to a cell gene and as a result activate it. Alternatively they can collect part of a cell's gene during this integration process, so that when the next generation of viruses are produced they have extra genetic material, and when this is inserted into the next cell it produces a transformation effect.

This mechanism for causing cancer has not been observed in humans. There is a group of retroviruses called human T-cell leukaemia viruses (HTLV) which are associated with rare types of leukaemia and lymphoma, but the gene that appears to cancer causing in these viruses does not appear to be related to any known human gene. So it appears that RNA viruses are not common causes of malignant disease in humans.

HTLV infection is endemic in localised regions in southern Japan, certain Caribbean islands, central America and northern Iran. Generally infection by this virus does not produce any symptoms of note. However, in a minority of cases two major clinical diseases may occur. One of these is tropical spastic paraparesis and the other is adult T-cell leukaemia/lymphoma (ATLL). All cases of ATLL have been observed to be associated with HTLV, where the viral genome has been found to be integrated in the host cells chromosomes.

By itself, infection by HTLV is insufficient to transform the cells and other mutations have been implicated. For example, the ATLL cells often have chromosome translocations affecting the region of chromosome 14 that contains the T-cell receptor gene. If this promoter region of this gene is brought into close contact with another gene, as a result of a translocation, then it may well activate that gene. Mutations in the *P53* gene have also been reported in ATLL cells.

The development of a vaccine against HTLV is not envisaged in the forseeable future.

Chapter 15
CONCLUSIONS

Introduction

It has been estimated that 30% of all cancers are primarily caused by an unhealthy diet, 30% are caused by tobacco and 20% by viral infection. This indicates that maybe as many as 80% of all cancers have a prime cause that can be eradicated by lifestyle choice.

Although these environmental influences are very important, the presence of genes which predispose to the development of cancer need always to be considered. Therefore the first stage in protecting yourself against cancer is self-awareness. Most cancers can be treated and the patient cured of the disease if the tumour is noticed early enough. With the advent of molecular genetic testing, which can reveal whether an individual carries cancer-predisposing forms of genes, a useful early warning of future problems can be provided. It is possible that this could alert an individual to potential future problems. In some cases this will mean actively watching for symptoms and following the best lifestyle choices in terms of diet and not smoking. In other cases where the predisposition may be very strong, individuals may in addition elect for more radical decisions, such as the presence of variant forms of *BRCA* genes which could lead some women to choose bilateral mastectomies.

Even where there is no clear genetic predisposition, individuals are always advised to be aware of possible symptoms. This is particularly the case with testicular cancer in men and breast cancer in women, where routine self-examination can reveal early changes. Needless to say, if a person notices any unusual symptoms they should not delay in contacting their doctors. The longer the tumour cells exist in the body, the longer they have to accumulate further changes which could make them impossible to treat.

It is one of the most unfortunate aspects of cancer that so many people are afraid of the disease that they would rather put off seeing their doctor when symptoms appear, in the hope that the symptoms will go away. If the symptoms are caused by cancer then this is very foolish because very few cancers spontaneously disappear. If the symptoms are not caused by cancer, then their doctor will be able to reassure the patient. That's part of their job. So there is no need to delay seeing a doctor about these concerns.

The Main Risks for the Most Common Cancers

Table 15.1 summarises the various risk factors that are thought to increase the chances of developing particular types of cancer, as well as factors which appear to be protective against cancer. Not all risks have the same potential in influencing the onset of malignant disease and in some ways it is difficult to precisely quantify this risk for each factor because of the complicated nature of cancer and because risks can interact to create an even bigger risk. However, we can be reasonably confident of predicting risk by placing these factors into high, moderate and low risk groups. If a risk factor occurs in a high risk group then it does not guarantee that you will develop that type of cancer, but it is highly likely. Conversely, if there is low risk factor, it does not necessarily mean that you can ignore it. The fact that it has been perceived as a risk means it has been associated with the development of cancer in some people and you could be such an at-risk individual.

Obviously the more risk factors that you identify as being relevant to your own life, the greater the chance of you developing cancer. Suppose we consider a worse case scenario and there is a woman who has had several female blood relatives who have died from breast or ovarian cancer. We know that her risks of developing a similar malignancy is 1 in 3, which is a very high incidence. What should she do? By examining the risk factors associated with breast cancer it is possible to design a lifestyle which reduces the risks of developing breast cancer. In other words, she should keep her weight to a healthy level (see Figure 12.1), she should not drink alcohol or smoke, she should have a diet high in fruit and vegetables and she should breast feed if possible.

In this book I have not been talking about guarantees, I have been talking about chances. Many people can tell of Aunt Doris who was a slim vegetarian and did not smoke and yet she still died in her 50s of breast cancer. They also may have had an Uncle Bob who smoked like a chimney, drank like a fish and died when he staggered out of a bar and into the path of a bus, at the age of 95. Anecdotes of this nature are meaningless. When you consider the chances of your developing cancer then you need to be aware of all the risks that likely to effect **you**. Close relatives may give you a clue as to how much you are at risk, because you may share similar genes but, beyond that, environmental factors are very important. As we still do not understand all of the interacting influences that affect cancer risk, it makes sense to pay particular attention to the ones we do know about.

Table 15.1 Risk factors and protective influences affecting cancer incidence.

Cancer type	Cases per year (worldwide)	Risk factor	Degree of effect	Protective influence
Lung	1,300,000	Smoking tobacco Occupational exposure to chemicals (eg., asbestos)	High	High fruit and vegetable consumption
		High dietary fat Alcohol consumption Radon gas exposure	Moderate Low	Regular physical activity High dietary vitamin C and E High dietary selenium
Stomach	1,000,000	High dietary salt	High	High fruit and vegetable consumption
		High smoked meat and fish consumption Bacterial (*Heliobacter pylori*) infection	Moderate Low	High dietary vitamin C High dietary carotenoids High dietary whole grain cereal Regular green tea consumption
Breast	900,000	Rapid growth in infancy and adolescence High body weight Alcohol	High Moderate	High fruit and vegetable consumption
		High dietary fat High red meat consumption Exposure to ionising radiation Late age at first pregnancy Smoking tobacco	Low	Regular physical exercise High dietary NSP (fibre) High dietary carotenoids Breast feeding
Colon and rectum	880,000	High red meat consumption Alcohol	High Moderate	Regular physical activity High vegetable consumption
		High body weight High dietary fat High egg consumption	Low	High dietary NSP (fibre) High dietary carotenoids Aspirin

147

Table 15.1 Continued.

Cancer type	Cases per year (worldwide)	Risk factor	Degree of effect	Protective influence
Mouth and throat	580,000	High dietary sugar Smoking tobacco Infestation by *Schistosoma sinensis* High alcohol consumption Chewing tobacco Smoking tobacco Regular consumption of very hot drinks	High Moderate Low	High fruit and vegetable consumption High dietary vitamin C
Liver	540,000	High alcohol consumption Hepatitis B virus infection Aflatoxin contaminated food	High Moderate Low	High vegetable consumption
Cervix	530,000	Papilloma virus infection Smoking tobacco	High Moderate Low	High fruit and vegetable consumption High dietary vitamin C and E High dietary carotenoids
Oesophagus	480,000	High alcohol consumption Smoking tobacco Regular consumption of very hot drinks	High Moderate Low	High fruit and vegetable consumption High dietary vitamin C High dietary carotenoids
Prostate	400,000	High dietary fat High red meat consumption High dairy products consumption	High Moderate Low	High vegetable consumption

Table 15.1 Continued.

Cancer type	Cases per year (worldwide)	Risk factor	Degree of effect	Protective influence
Bladder	310,000	Occupational exposure to chemicals	High	High fruit and vegetable consumption
		Smoking tobacco	Moderate	
		Infestation by *Schistosoma haematobium*	Low	
Pancreas	200,000	Smoking tobacco	High	High fruit and vegetable consumption
			Moderate	
		High energy intake	Low	High dietary NSP (fibre)
		High dietary cholesterol		High dietary vitamin C
		High red meat consumption		
Ovary	190,000		High	High fruit and vegetable consumption
			Moderate	Extended use of oral contraception
		No childbirth	Low	

The Key to Increasing your Chances of a Healthy Cancer-free Life

Below is your lucky 13 check list of ways in which you can greatly reduce your risks of dying from cancer.

Vigilance

1. The first priority in stacking the odds in your favour is self-awareness. A conscious effort to note any changes in your body, which might indicate the early development of a tumour should be referred to a doctor. This will greatly enhance your chances of a cure if a cancer does occur. Of particular relevance are the occurrence of lumps, persistent bleeding, a change in a mole, persistent cough, persistent hoarseness, a change in bowel habits or an unexplained weight loss.
2. It is also important to take advantage of screening programmes that may be available, such as Pap smears and mammograms.

Behaviour

3. Do not smoke, or use any form of tobacco, at all.
4. Take exercise regularly and keep your weight within healthy levels (see Chapter 12).
5. Sexual activity with different partners should be restricted and, where it occurs outside of a stable relationship, should involve the use of condoms.
6. Use vaccinations where appropriate. Vaccinations are currently available against hepatitis B virus and hopefully in the future will be available against human papilloma viruses.
7. Avoid sunburn, particularly if you are white and a child.
8. Follow Health and Safety instructions at work concerning the production and handling of any substances that have a known cancer risk.

Diet

9. Keep alcohol consumption to no more than 1–2 units per day for men and no more than 1 unit per day for a female. This limit is particularly important if you also smoke.

10. Choose a diet which is primarily vegetarian and consume at least 600g of fruit and vegetables each day.

11. Red meat from domesticated animals should be eaten only occasionally with an average daily consumption of less than 80g. Poultry and fish are preferable meats.

12. Fat consumption should be limited to provide between 15–30% of total energy requirement. Fats from animal origins should be avoided where possible.

13. Reduce salt content in the diet to less than 3g per day and only occasionally eat highly salted processed foods (crisps, bacon), salt-preserved food (salted dried fish), or pickled vegetables.

INDEX